Semiconductor Integrated Optics for Switching Light

Media content is available from the book information online: https://doi.org/10.1088/978-1-6817-4521-3.

ISBN   978-1-6817-4521-3 (ebook)
ISBN   978-1-6817-4520-6 (print)
ISBN   978-1-6817-4523-7 (mobi)

DOI   10.1088/978-1-6817-4521-3

Version: 20170801

IOP Concise Physics
ISSN 2053-2571 (online)
ISSN 2054-7307 (print)

A Morgan & Claypool publication as part of IOP Concise Physics
Published by Morgan & Claypool Publishers, 40 Oak Drive, San Rafael, CA, 94903 USA

IOP Publishing, Temple Circus, Temple Way, Bristol BS1 6HG, UK

Contents

Preface

This book has its genesis in research work on all-optical switching in semiconductors. After sifting through many types of materials and device configurations the author and his collaborators were guided by the theory of optical waveguiding, two-photon absorption and the strongly related theory of the pure optical Kerr effect in semiconductors and arrived at semiconductor waveguide devices for ultrafast switching. There is a strong emphasis on III–V semiconductors. In particular for optical communications wavelengths it turns out that one of the best materials is based on AlGaAs semiconductor alloys. Specifically, at for example optical communications wavelengths in the so-called C band around 1550 nm, then with Al fraction in Al_xGa_{1-x} As around $x \sim 0.18$ in the core of semiconductor waveguides, then this leads to the best semiconductor devices for ultrafast all-optical switching. Much of the book is dedicated to explaining why this is the case and discussing the evidence for this conclusion. Special thanks go to Professor J Stewart Aitchison, now of the University of Toronto, who came with idea of using the $Al_{0.18}Ga_{0.82}As$ alloy.

There are some interesting aspects of the effect of electric fields on the optical properties of semiconductors covered along the way. These include the linear electro-optic effect, electroabsorption and electrorefraction and how these relate to the second and third order nonlinear optics of semiconductors. The cascaded second order nonlinear optical effect, associate with linear electro-optic effect, can lead to a nonlinear refractive index normally associated with the third order nonlinear optical effect. Further, the theory of two-photon gain in semiconductors is derived by combining simple semiconductor laser theory with two-photon absorption theory. Two photon gain in semiconductors has now been observed by some research groups.

In order to give insight into how all the above theory is combined and implemented in the design of devices, several Mathematica programs are included. These can be downloaded from the book information online: https://doi.org/10.1088/978-1-6817-4521-3.

The author hopes that the reader finds the same amount enjoyment in reading about this work as the author and his collaborators (see the acknowledgements) found in doing the work.

Acknowledgements

Without outstandingly talented, collaborative and supportive research colleagues the work covered in this book would not have been possible. In research colleagues I include students, academic staff, administrative staff, technicians, staff from funding bodies and industrial colleagues. It is too long a list of people to name (check out the references!), but one late colleague Professor George Stegeman (1942–2015) of the University of Central Florida deserves particular mention; his research group made many important contributions to this field leaving a legacy that continues to inspire a new generation of research workers.

Author biography

Charlie Ironside

Professsor Charlie Ironside, BSc, PhD, FIET, FInstP has over 30 years' experience in semiconductor optoelectronics research and in particular microfabrication of semiconductor photonic components such as nonlinear optical switches, laser diodes and components for optical sensing and optical metrology systems. He has published over 125 research journal publications and 140 conference papers. He has co-authored five patents and has won an award for transferring knowledge from university research to industry for commercial exploitation. He is a Fellow of the institution of Engineering, Technology (FIET), a Fellow of the Institute of Physics (FInstP) and senior member of the Institute of Electronics and Electrical Engineering (IEEE).

He did his PhD work at Heriot-Watt University, Edinburgh, UK, and then moved to do post-doctoral research at the University of Oxford. In 1984 he moved from physics at Oxford to engineering at the University of Glasgow. In 1998 he was appointed Professor of Quantum Electronics, in the Department of Electronics and Electrical Engineering at the University of Glasgow. In 2014 he moved to the Department of Physics and Astronomy, Curtin University in Western Australia where he is currently a professor.

His research highlights include: fastest all-optical semiconductor switch, invention of the multiple colliding pulse passively mode-locked laser diode, invention of the two-photon semiconductor laser, the first quantum cascade laser made in Europe and supervision of the first PhD on quantum cascade lasers, pioneering of resonant tunnelling diode optoelectronic integrated circuits, grating microtraps for cooling atomic vapour to micro kelvin temperatures.

For more details see: http://oasisapps.curtin.edu.au/staff/profile/view/Charlie. Ironside.

IOP Concise Physics

Semiconductor Integrated Optics for Switching Light

Charlie Ironside

Chapter 1

Introduction

1.1 Introduction

Taking a broad perspective of information technology systems, switching is one of the key functions along with communications and storage. In this book we are updating, revising, expanding and refreshing a review of all-optical switching from 1993 [1] to make it relevant for readers who are interested in a review of this very active area of physics and engineering research [2].

Improvements to switching technology, particularly increasing the speed of operation and reducing the energy required, will provide immediate benefits in information technology systems. Currently, the set-up of optical communications packet switching of data under the TCP/IP (Transmission Control Protocol (TCP) and the Internet Protocol (IP)) protocols requires that the address is read from a packet and the address information is used to configure the switches in a router. Switching is also the fundamental operation required in digital signal processing and computing.

1.2 Switching

The broad landscape of switching technology for optical communication systems capable of operating at data rates of 100 GB s^{-1} is summarized in figure 1.1 (adapted from [3]). Two key drivers of the technology are plotted: the switching energy required per bit and the length of the switch. These are closely related to the operating cost of an optical communications system that is the power required to run it and the footprint of the switching technology. It is apparent that the sweet spot in this diagram is the bottom left hand corner.

This is a very active research area and figure 1.1 does not attempt to be fully comprehensive, rather just representative of some promising approaches. Further, as adapted, figure 1.1 is only to be regarded as a rough map of the landscape; more detail can be obtained from [3]. It should be borne in mind that the metrics chosen to define this landscape, switching energy and length of switch, while useful in most

doi:10.1088/978-1-6817-4521-3ch1

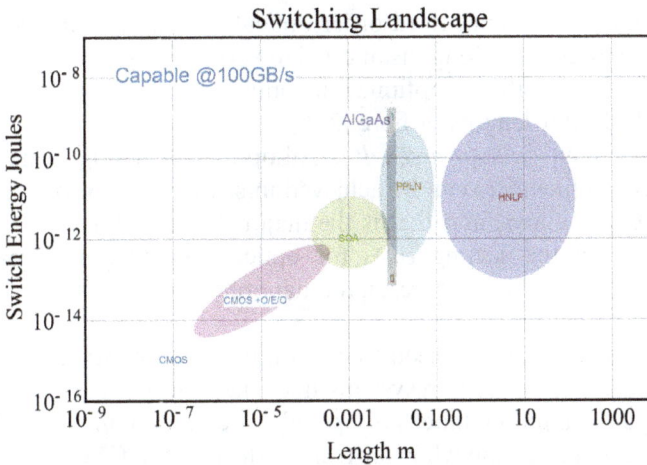

Figure 1.1. Adapted from Hinton *et al* [3], the energy per bit versus device length for photonic signal processing technologies. AlGaAs refers to the all-optical switching in AlGaAs optical waveguides, the main topic of this book. CMOS refers to the major digital electronics switching technology, including O/E/O optical to electrical and electrical to optical conversion. SOA refers to semiconductor optical amplifiers; PPLN refers to periodically poled lithium niobate; HNLF refers to highly nonlinear fibre. The colour areas around the acronyms are approximately indicative of the part of the switching landscape occupied by these technologies.

respects, are not the only metrics that could be applied. For example, implicit in the switching energy metric is the assumption that the energy is consumed by the switch and turned into heat, and that is not the case for some of the all-optical switches.

CMOS (complementary metal oxide semiconductor). Technology consisting of MOSFETs (metal oxide field effect transistors) is the dominant switching technology and the *de facto* benchmark. CMOS is entirely an electronic switching technology so the optical to electrical and electrical to optical (O/E/O) part of the system has to be included if we are to make a fair comparison with the all-optical technologies.

SOAs (semiconductor optical amplifiers). An SOA can be employed as an all-optical switch technology. An SOA is a semiconductor laser configured to operate as an optical amplifier. When the current is injected in addition to optical amplification it has large nonlinear optical effects that can be used to all-optically switch [4].

PPLN (periodically poled lithium niobate). PPLN can be employed as an all-optical switch technology. Lithium niobate is routinely used for its second order nonlinear optical effects leading to three-wave mixing devices such as second harmonic generators. This material has a well-developed technology for phase matching in second order devices that uses a regular spatial variation in the lithium niobate crystal consisting of a periodic sign change in the second order nonlinear susceptibility of the medium—this is called periodic polling. A similar concept in the context of AlGaAs waveguides is discussed in section 3.2. See also [5].

HNLF (highly nonlinear fibre). All-optical switching in optical fibre was the first guided wave all-optical switching technology. Optical fibre devices made use of the outstanding transparency (~ 0.2 db km^{-1} at wavelength around 1550 nm) of the fibre to make use of the small third order optical nonlinearity to acquire optically induced phase changes over long distances. However, by using the theory described in section 3.3 it was possible to optimize the optical nonlinearity and design HNLF that used much shorter lengths of fibre [2, 6].

AlGaAs (aluminum gallium arsenide). All-optical switching using both second and third order nonlinearites can be achieved in semiconductor optical waveguides made from AlGaAs alloys, and that is the major theme of this book.

Another take on the scaling rules for optical switching with nanophotonics and generally concentrating on resonant nanophotonic structures can be found in [7].

Using one signal, usually in electronic form, to switch another signal has been a key building block of information systems since their inception, and at first glance it seems as though the building blocks of optical switching are in place. However, it is hard to improve on perfection while far from perfect silicon CMOS (complementary metal oxide semiconductor) based on MOSFETs (metal oxide semiconductor field effect transistors) technology for digital switching has swept all before it since its introduction in the 1960s.

If CMOS does have any weakness it may be in the heat management associated with ultrafast switching. The CMOS industry has made great strides in advancing the capability of CMOS devices but it appears that the thermal diffusion coefficient of silicon, $D_T = 9 \times 10^{-5} \, \text{m}^2 \, \text{s}^{-1}$, is something the industry is stuck with (unless it ventures into diamond $D_T = 1.2 \times 10^{-3} \, \text{m}^2 \, \text{s}^{-1}$). So the power density associated with a MOSFET switching operation is more or less constant according to Denard scaling [8] (although that may have been broken in recent years as the gates are now below 100 nm length). It is fundamental to information processing, since there is a lower limit to how much energy is required to represent 1 bit of information, that energy limit is set by the inherent sources of noise present in any physical system that is employed to represent and process information. Currently the limit for CMOS is in the <1 fJ (figure 1.1). As the processing speeds increase, then all those femtojoules add up and power required for information processing increases linearly with bit rate. The bottom line is that, for example in microprocessors, the speed of operation as determined by the clock frequency has been stuck around 3–5 GHz since around 2004 [9]. One of the last big improvements came when copper with higher electrical conductivity replaced aluminum as the electrical conductor used in electrical interconnects [10].

So CMOS switching is limited by thermo-dynamics and, indeed, in large server farms the plumbing, associated with heat management, is almost as important as the electronics [11]. The thermo-dynamics of switching, determined by the second law, is such that every switching operation results in an increase in entropy. In CMOS that entropy ends up as heating, however, in some forms of all-optical switching there is still a price to be paid in terms of entropy but it does not end-up in heat and that brings some advantage to ultrafast all-optical switching.

All-optical switching looks like an elegant engineering solution to switching in optical communication networks partly because it eliminates the overhead associated with optical to electrical and then electrical to optical (OEO). However, as reviewed in [3], even with the OEO overhead, CMOS is such a mature technology with a vast legacy of investment that it is hard to see how all-optical switching will displace the incumbent technology unless an entirely new paradigm is adopted for switching information.

In most considerations of optical properties of materials it is assumed that the optical properties of the materials are independent of the intensity of the light used to measure the optical properties. But for sufficiently high intensities it is possible to alter the optical properties of materials. There are many physical mechanisms that can make the optical properties of materials intensity dependent, including thermal effects via optical absorption and molecular re-orientation, and for a comprehensive review see Boyd [12].

In the context of fast switching of light, in this book we are concerned with optical nonlinearities arising from the application of electric fields and the effect of these fields on the optical properties of the material, usually through the effect the electric field has on the bound electrons, that is, electrons involved in bonding the material and not free to carry charge. However, there has been a great deal of effort in successfully employing free carriers in all-optical switching [2] and in using resonant structures to store light and increase the local optical intensity [7]. But in the spirit of exploring the ultimate, switching, speed limit this book concentrates on nonlinearities associated with bound electrons in non-resonant devices.

1.3 Linear optics: response theory and the Kramers–Kronig transformation

Starting with linear optics, we go on to describe how applied electric fields alter the optical properties of the material. It turns out that electric field can be applied either via electrodes attached to the material or via the light itself. The usual way of dealing with these phenomena is to modify and extend the concepts introduced by Maxwell in his set of equations that describe the propagation of electromagnetic radiation as a polarizable medium, a dielectric, if there are no free charge carriers.

Maxwell [13] postulated that the electric field E of the light field induces a polarization P in the medium and that

$$P = \chi(\omega)E(\omega), \tag{1.1}$$

where $\chi(\omega)$, is a frequency-dependent variable material parameter known as the electric susceptibility. This is a classic linear systems response theory. So the material responds to the application of an electric field and the net result is a polarisation, $P(\omega)$, and the material response is represented by the susceptibility, $\chi(\omega)$, that is dependent of the frequency, ω, of the applied electric field, $E(\omega)$. In general, $\chi(\omega)$, is a complex number with the real part giving rise to refractive index and the imaginary part giving rise to absorption. This treatment is what became much later than Maxwell known as a linear systems treatment. So the material is regarded as a system that responds to an applied electric field by producing a polarization, P and the material response is frequency dependent, that is, in general we can expect the response of the material to depend on the frequency of the applied electric field—that is because it will take a finite amount of time for the material to respond to changes in electric field which may sound obvious but it did elude Maxwell.

Now that linear systems theory is well developed we can borrow results from linear systems. So for example equation (1.1) can be transformed into the time domain by applying the convolution theorem which states that taking the inverse Fourier transform of the product of two functions in the frequency domain is equal to the convolution of the inverse Fourier transforms of the functions. This can be more neatly expressed as

$$P = \chi(\omega)E(\omega) \overset{\text{Fourier Pair}}{\Longleftrightarrow} \chi(t) \otimes E(t). \tag{1.2}$$

The convolution theorem can be applied to this analysis to provide further insight. We explicitly include the time dependence of the polarization to get

$$p(t) = \chi(t) \otimes E(t). \tag{1.3}$$

Rewriting in full integral form gives

$$p(t) = \int_{-\infty}^{\infty} \chi(t)E(t - \tau)d\tau, \tag{1.4}$$

where τ is the dummy variable and, $\chi(t)$, is the impulse response of the electric susceptibility; it is the material response to the ultimate ultrashort pulse.

In the original Maxwell theory, because he was unaware of the exact nature of matter in as much as he was unaware of the existence of the electron, it is implicitly assumed that the material responds instantaneously to the applied electric field and that is why the frequency dependence of the polarisation eluded Maxwell. However, because the material does not response instantly there is a frequency dependence or dispersion of the polarisation. Moreover, Maxwell's approach does involve a degree of averaging over a volume of the material. Even within entirely classical physics the exact nature of the material response to electric field and electromagnetic wave propagation and the relationship between them has approximately kept the large-scale averaging with correction for local field effects [14].

There is a relationship between the dispersion of the absorptive and refractive nonlinear optical effect, electroabsorption and electrorefraction, which plays a very important role in the application of these effects and can be traced back to the principle of causality, which states that the effect cannot precede the cause. So the medium cannot respond, the polarisation does not occur, before the electric field is present.

In the time domain we use the unit step function, $u(t)$, to implement causality where

$$\begin{aligned} u(t) &= 0 \text{ for } t < 0 \\ &= 1 \text{ for } t > 0. \end{aligned} \tag{1.5}$$

So that with causality, $\chi(t)$ *becomes* $u(t)\chi(t)$, and we can now apply the convolution theorem ($\otimes$, symbol for convolution) to see how this comes over to the frequency domain

$$\widehat{\chi(\omega)} = \chi(\omega) \otimes \left(\frac{\delta(\omega)}{2} + \frac{i}{2\pi\omega} \right)$$

$$= \chi(\omega) + \frac{i}{2\pi}\mathcal{P} \int_{-\infty}^{\infty} \frac{\chi(\Omega)}{\Omega - \omega} d\Omega \qquad (1.6)$$

$$= \frac{i}{\pi}\mathcal{P} \int_{-\infty}^{\infty} \frac{\chi(\Omega)}{\Omega - \omega} d\Omega.$$

In linear systems theory this is Hilbert transform—the arrow, $\widehat{\chi(\omega)} \rightarrow \chi(\omega)$ indicates that it is an analytic function. To make use of the Hilbert transform in our present context we need to take account of exactly what is meant be causality. In some ways it is surprising that causality actually applies to light because photons travel at the speed of light and therefore according to special relativity they do not experience time or space—in the photon frame of reference it is created and annihilated in the same instant. By a careful consideration of causality Kramers and Kronig adapted the Hilbert transform to work for electromagnetic waves—for comprehensive reviews see [15, 16], and for a very much shorter review see [17].

1.4 Nonlinear optics

Nonlinear optics, in the sense of an electric field altering optical properties, is almost as old as Maxwell's electro-magnetic theory of light and indeed some of the very early experiments reported in 1875 [18] by John Kerr were carried out partly to confirm the electromagnetic character of light. The Pockels effect is related to the Kerr effect, first reported in 1893. The Kerr effect is the change of the refractive index varying as the square of the electric field and with Pockels effect the refractive index varies linearly with electric field. As we shall see, these effects are related to the second order or three-wave mixing effects (Pockels effect) and the third order or four-wave mixing effect (Kerr effect). The reason the Kerr effect was discovered first is related to the fact that it occurs in all materials—the magnitude varies a lot but it is present in all materials. The Pockels effect requires a crystalline material that lacks a centre of symmetry.

Usually the Kerr and Pockels effects modulators use a quasi-static electric field to modulate light. So the quasi static electric field varies with a frequency much less than the electric field of the light which typically has a frequency of around 10^{14} Hz. But the electric field of the light can alter the susceptibility, $\chi(t)$, of the medium $P = \chi(t) \otimes E(t)$ then it is possible for electric field of the light itself to alter the polarisation.

Insight into both the effects can be obtained from the polynomial series expansion of polarisability (there are many books that cover nonlinear optics, see for example Boyd [12]),

$$p = \chi^{(1)}E + \chi^{(2)}EE + \chi^{(3)}EEE... \qquad (1.7)$$

of a medium and this is the standard treatment used in nonlinear optics [19]. This way of describing the nonlinear effects ties in very neatly with Maxwell's equation and the 'dielectrification' of the medium that Kerr was keen to observe, providing evidence to validate Maxwell's equations.

The second order nonlinear term, $\chi^{(2)}EE$, is associated with three-wave mixing and gives rise to processes such as second harmonic generation. The third order term, $\chi^{(3)}EEE$, is associated with four-wave mixing and gives rise to processes such as two-photon absorption and the nonlinear refractive index.

The difference between the Pockels effect and the Kerr effect, where in the Pockels effect the change in refractive index is linearly proportional to the change in the electric field and with the Kerr effect the change in the refractive index is proportional to the square of electric field, means that, with the Pockels effect if the electric field is reversed then the sign of the change in the refractive index is reversed but that does not happen with the Kerr effect.

Kerr measured the change in the optical properties via the change in the real part of the refractive index that made the material birefringent and thus rotated the plane of the polarisation of the light. But that the medium would also have changed its absorption spectrum at some wavelength that was not apparent to Kerr, this occurs because of the Kramers–Kronig relationship between the real and imaginary parts of the refractive index.

Recently, there has been a great deal of progress in quantum optics where nonlinear optical effects are enhanced to the point where the interaction between single photons can be observed and utilised—see for example [20]. A map of nonlinear interactions is shown in figure 1.2. While the ultimate in the strength of nonlinear optical effects is the quantum photon–photon nonlinear optics and quantum many body regions, the media required are delicate atomic vapours and the timescales are not ultrafast because the interaction generally requires resonant structures that store light on long timescales. So the region of the map we are concerned with in this book is the classical nonlinear optics region where we utilise relatively weak nonlinear effects and with relatively high intensity ultrashort (~1 ps) pulses of light.

1.5 Integrated optics—semiconductor optical waveguides

In bulk optics where a lens is used to focus light to obtain high intensity in a material, the length of the region subject to high intensity is limited by diffraction. To avoid the limit imposed by diffraction, all the practical all-optical switching devices have employed optical waveguides to confine the high intensity light in lengths not limited by diffraction but by loss, either absorption or scattering losses. In this book we are concerned with integrate optics and waveguides fabricated on semiconductor chips. However, there has also been a great deal of success with all-optical switching devices in optical fibre formats, see for example [21].

For several reasons optical waveguides have been developed for semiconductor optoelectronics [22]. Primarily, they were developed for semiconductor lasers which only became practical devices because of semiconductor heterostructures that consisted of epitaxial layers of semiconductor alloys on a semiconductor substrate.

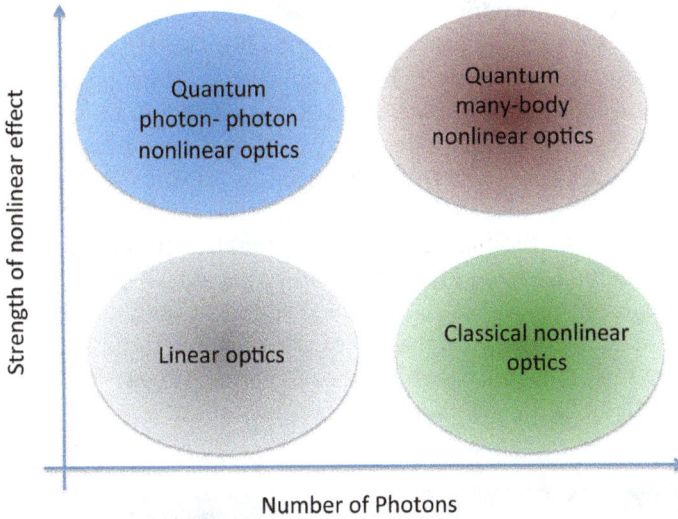

Figure 1.2. Adapted from [20], is a schematic map of nonlinear optical interactions. In this book we are concentrating on the region designated classical nonlinear optics accessible with semiconductor optical nonlinearities in AlGaAs integrated optics.

In the heterostructure semiconductor laser a narrow band-gap semiconductor alloy is sandwiched between lower band-gap semiconductors. In the classical case of the AlGaAs/GaAs material system $Al_xGa_{1-x}As/Al_yGa_{1-y}As/Al_xGa_{1-x}As$ forms layers where $x > y$ resulting in a heterostructure that confines electrons, holes and light to the $Al_yGa_{1-y}As$ region. The key to the optical waveguiding is that the small band-gap material also has the largest refractive index and through total internal reflection the light is confined to the small band-gap region. The light can further be confined in the lateral direction by a mesa structure on the surface of the chip.

The key advantage of waveguide formats for nonlinear optics is that high intensity can be maintained over long lengths. In free space the distance over which high intensity can be maintained is limited by diffraction, but in waveguide formats that distance is limited only by linear absorption, nonlinear absorption and scattering losses. The high-intensity interaction lengths required for efficient non-linear optics processes are much more easily achieved with a waveguide format.

For electroabsorption and electrorefraction the advantage of the waveguide format is that light is confined to a small region of the chip so that a small voltage (∼5 V) can be used to produce a large electric field of several kV cm^{-1} in the region of the semiconductor chip where the light is confined.

The semiconductor alloy $Al_xGa_{1-x}As$ is the foundation material for semiconductor heterostructures [23–25]. It can been grown in epitaxial layers by a variety of techniques including molecular beam epitaxy (MBE), and is extensively employed in optoelectronic devices such as the light emitting diode (LED) and the semiconductor laser. In this book we will extensively employ AlGaAs to illustrate device concepts but the device concepts we develop do have a wider range of applications.

Figure 1.3 shows a schematic of a generic AlGaAs semiconductor waveguide. A simple effective index method (see for example [27]) can be used to determine the waveguide modes, which is how light is confined in the waveguide. A scanning electron micrograph (SEM) of AlGaAs waveguide is shown in figure 1.4.

In first pass design the effective index is useful in determining the required layer thickness, strip width and refractive indices of the various layers making up the waveguide.

The effective index method gives a first approximation to the type of optical mode that will propagate in a semiconductor optical waveguide. The result of an effective index method calculation for an AlGaAs waveguide is shown in figure 1.5.

A much more detailed analysis that takes into account the full vectorial nature of the light is found in [28]. This paper has an associated website from which software can be downloaded [29] to calculate the waveguide modes.

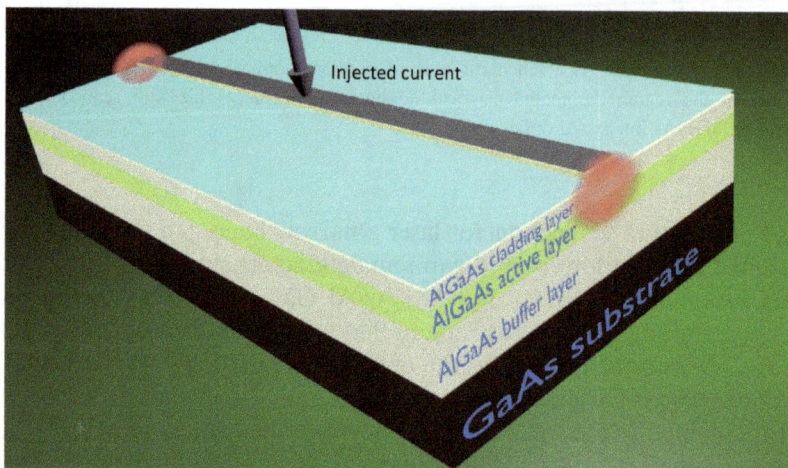

Figure 1.3. Schematic of the generic AlGAs waveguide showing a semiconductor laser layout.

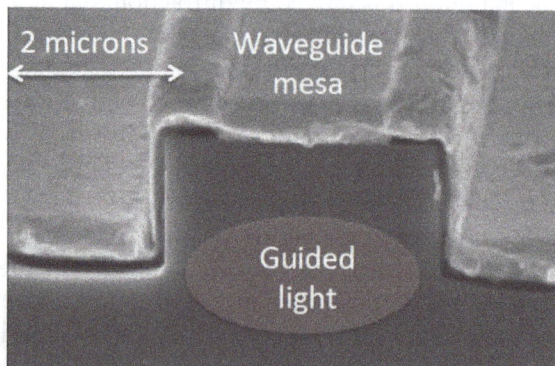

Figure 1.4. Scanning electron micrograph (SEM) of the end facet view of AlGaAs waveguides; indicated is the region at the end facet where the guided light is confined [26]. This is part of a semiconductor laser.

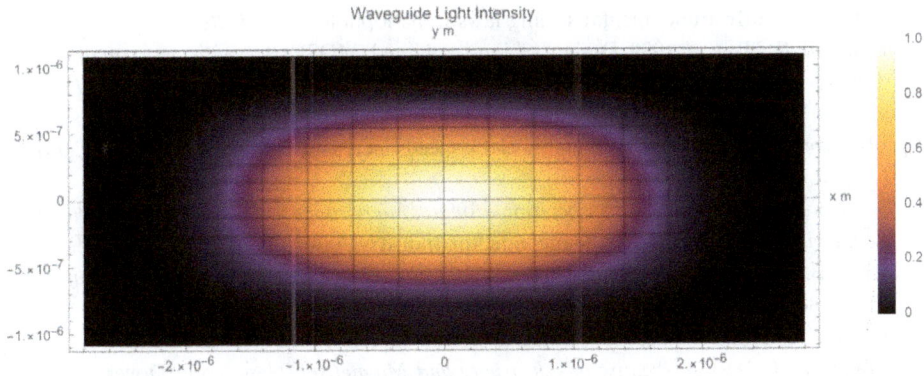

Figure 1.5. The figure shows a two-dimensional plot of the normalised light intensity distribution, calculated using the effective index method, of the optical mode a semiconductor optical waveguide. Using the parameters of the AlGaAs waveguide described in figure 4.4. See the Mathematica notebook Waveguide Effective Index Channel.nb.

1.6 Outline of book

In this book we will concentrate on AlGaAs semiconductor integrated optical devices for ultrafast (~1 ps) all-optical switching. As an introduction we put the AlGaAs work context in chapter 1 by a review of the research landscape associated with devices appropriate for future development of high data rate (~100 GB s^{-1}) optical communication systems. Further, in chapter 1 we cover some general topics required in future chapters such as linear optics response theory, nonlinear optics integrated optics and optical waveguiding. In chapter 2 we cover aspects of the linear electro-optic effect, electroabsorption and electrorefraction required for electrical/optical conversion and also as way of introducing nonlinear optical effects. Chapter 3 discusses the nonlinear effects in AlGaAs waveguides, cascaded second order nonlinearity and the optimized third order nonlinearity that can be used in the integrated optic devices for all-optical switching discussed in chapter 4, the cascade push–pull Mach–Zehnder interferometer, the nonlinear coupler and the asymmetric Mach–Zehnder interferometer. Chapter 5 presents conclusions and future directions.

References

[1] Ironside C N 1993 Ultra-fast all-optical switching *Contemp. Phys.* **34** 1–18

[2] Wabnitz S and Eggleton B J 2015 *All-Optical Signal Processing* (Springer Series in Optical Sciences vol 194) (Berlin: Springer)

[3] Hinton K *et al* 2008 Switching energy and device size limits on digital photonic signal processing technologies *IEEE J. Sel. Top. Quantum Electron.* **14** 938–45

[4] Bonk R 2013 *Linear and Nonlinear Semiconductor Optical Amplifiers for Next-Generation Optical Networks* (Karlsruhe: KIT Scientific Publishing)

[5] Langrock C *et al* 2006 All-optical signal processing using $\chi^{(2)}$ nonlinearities in guided-wave devices *J. Lightwave Technol.* **24** 2579–92

[6] Asobe M, Kanamori T and Kubodera K 1992 Ultrafast all-optical switching using highly nonlinear chalcogenide glass-fiber *IEEE Photonics Technol. Lett.* **4** 362–5

[7] Liu K *et al* 2016 Fundamental scaling laws in nanophotonics *Sci. Rep.* **6** 12

[8] Dennard R H *et al* 1999 Design of ion-implanted MOSFET's with very small physical dimensions (Reprinted from 1974 *IEEE J. Solid-State Circuits* **9** 256–68) *Proc. IEEE* **87** 668–78

[9] Mattsson P P 2014 Why haven't CPU clock speeds increased in the last few years? *COMSOL blog* Nov. 13, 2014

[10] Semiconductor Industry Association 1999 *International Technology Roadmap for Semiconductors* (Austin, TX: Semiconductor Industry Association) pp 163–86

[11] Ristic B, Madani K and Makuch Z 2015 The water footprint of data centers *Sustainability* **7** 11260–84

[12] Boyd R W 2008 *Nonlinear Optics* (Amsterdam: Elsevier)

[13] Maxwell J C 1891 *A Treatise on Electricity and Magnetism* (New York: Dover)

[14] Aspnes D E 1982 Local-field effects and effective-medium theory—a microscopic perspective *Am. J. Phys.* **50** 704–9

[15] Lucarini V, Sarinen J J, Peiponen K-E and Vartiainen E M 2004 *Kramers–Kronig Relations in Optical Materials Research* (Springer Series in Optical Sciences vol 110) (Berlin: Springer)

[16] Hutchings D C *et al* 1992 Kramers–Kronig relations in nonlinear optics *Opt. Quantum Electron.* **24** 1–30

[17] Hu B Y K 1989 Kramers–Kronig in 2 lines *Am. J. Phys.* **57** 821

[18] Kerr J 1875 A new relationship between electricity and light: dielectrified media birefringent *Phil. Mag.* **50** 337

[19] Butcher P N and Cotter D 1990 *The Elements of Nonlinear Optics* (Cambridge: Cambridge University Press)

[20] Chang D E, Vuletic V and Lukin M D 2014 Quantum nonlinear optics— photon by photon *Nat. Photon.* **8** 685–94

[21] Cotter D *et al* 1999 Nonlinear optics for high-speed digital information processing *Science* **286** 1523–8

[22] Tamir T, Griffel G and Bertoni H L 1994 Guided-wave optoelectronics: device character-ization, analysis, and design *Int. Symp. Proc.* (*Brooklyn, NY October* 26–28, 1994)

[23] Alferov Z I *et al* 1972 Recombination radiation emitted by 4-layer structures based on GaAs-AlAs heterojunctions *Sov. Phys. Semicond.-USSR* **6** 637

[24] Capasso F 1987 Band-gap engineering—from physics and materials to new semiconductor-devices *Science* **235** 172–6

[25] Orton J 2004 *The Story of Semiconductors* (Oxford: Oxford University Press)

[26] Tandoi G 2011 Monolithic high power mode locked GaAs/AlGaAs quantum well lasers *Thesis* University of Glasgow.

[27] Lee D L 1986 *Electromagentic Principles of Integrated Optics* (New York: Wiley)

[28] Fallahkhair A B, Li K S and Murphy T E 2008 Vector finite difference modesolver for anisotropic dielectric waveguides *J. Lightwave Technol.* **26** 1423–31

[29] Fallahkhair A B, Li K S and Murphy T E 2017 *WGMODES optical eigenmode solver for dielectric waveguides*, available from http://www.photonics.umd.edu/software/wgmodes/

Chapter 2

Linear electro-optic effect, electroabsorption and electrorefraction

2.1 Dielectricfication: linear electro-optic effect, electrorefraction and electroabsorption

In this chapter we cover the effect that the application of an electrical field has on the optical properties of semiconductors partly because this is important in electrical to optical data conversion and partly as an introduction to nonlinear optical effects.

Some of the very early experiments reported in 1875 [1] by John Kerr were carried out to confirm the electromagnetic character of light. The Pockels effect first reported in 1893 is related to the Kerr effect. The Kerr effect is the change of the refractive index varying as the square of the electric field, and with the Pockels effect the refractive index varies linearly with electric field. The Pockels effect is related to the second order nonlinear optical effect (or three-wave mixing effects and the Kerr effect is related to the third order nonlinear optical effect (or four-wave mixing effect). The Pockels effect or linear electro-optic effect (LEO) requires a crystalline material that lacks a centre of symmetry—so for example, while GaAs has a LEO effect, silicon which is more symmetric does not exhibit LEO.

Figure 2.1 shows a photograph of where nonlinear optics started, it is the original Kerr cell from 1875, still in the possession of the University of Glasgow, Hunterian museum. The Kerr cell is made of a glass material with two embedded electrodes. When a voltage is applied across the electrodes, an electric field is produced between the electrodes and the material becomes optically birefringent; that is, the light polarised parallel to the electric field between the electrodes has a different refractive index for light perpendicular to the electric field. Kerr measured a rotation of the angle of polarisation of the light passing through the glass because of the electrically induced birefringence. The gap and the electrodes (the metal rods) are clearly visible in the photograph shown in figure 2.1

doi:10.1088/978-1-6817-4521-3ch2

Figure 2.1. A photograph of the original Kerr cell shows two electrodes embedded in glass, the gap between the electrode rods is around 1 cm. The electric field between the electrodes altered the refractive index for light and made the glass birefringent. That is, when an electric field was applied, light with polarisation parallel to the axis of the electrodes experienced a different refractive index compared to light with polarisation perpendicular to the axis of the electrodes. From Kerr J 1875 A new relationship between electricity and light: dielectrified media birefringent *Philos. Mag.* **50** 337.

Just a note here about terminology: it would appear logical that the generic term for the effect of electric field on optical properties should be electro-optics, however, that is not the case; in the normally accepted terminology, electro-optics refers to changes in refractive index via the Pockels effect, the linear electro-optic effect (LEO). The changes in refractive index that come from the Kramers–Kronig transform of the absorption band-edge (in bulk semiconductors due to the Franz–Keldysh effect) is termed electrorefraction [2]. In Kerr's terminology, now archaic, dielectrification leads to both electrorefraction and electroabsorption. Also the electric field applied in the Kerr cell is relatively static compared to the electric of the light wave, which is oscillating typically in the region of 10^{14}Hz, whereas the applied electric field was essentially held constant in Kerr's experiments. The generic term used in this book to refer to static electric field induced refractive index changes will be electro-refractive effects, so that includes both LEO and electrorefraction.

The Kerr effect relates to the change in refractive index, however, as we can see from the Kramers–Kronig relationship, a change in refractive index implies a change in absorption coefficient. So in the glass cell, when the electric field is applied to the glass there is also a change in the absorption coefficient. Subsequent work revealed that the change in the absorption coefficient was mostly at photon energies close to the band-edge in bulk solids; this is the so-called Franz–Keldysh effect [3]. Kerr could not observe the band-edge change in absorption because for glass the band edge is around 9 eV and the visible photons available to Kerr have energies in the range 1–2 eV.

LEO and electroabsorption have been used extensively as the basis of integrated optical devices for modulating light and for digital optical communications [4, 5]. The devices are used to encode light from semiconductor lasers with digital information. Usually in non-return-to-zero (NRZ) formats the light is amplitude modulated, that is, simply turned on and off to represent binary data, 1 and 0. In digital communications the applied electric field is no longer static and can vary with a bandwidth up to $\sim 10^{12}$Hz.

At least conceptually, the LEO devices are easily converted into devices where the signal that induces the change in the refractive index is not an electronic signal but is a photonic signal. With the electronic signal the electrically induced refractive index change in the waveguide comes from a voltage applied to the electrodes on top of the waveguide that produces an electric field in the waveguide, and via the Kerr and or Pockels effect the refractive index of the waveguide is changed. With the photonic signal the electric field comes from light propagating in the waveguide inducing a refractive index change through nonlinear optical effects.

So this poses the question: what are the fundamental differences between applying the control signal as an electronic or a photonic signal?

In the electronic signal case we can regard the metal electrodes as waveguides and the metal has a large refractive index that guides the electronic signal. The electrons in the metal waveguide are fermions with mass and need to be moved around to create the charge imbalance that produces the electric field across the waveguide. Moving fermions that have rest mass usually involves evoking the terminology used in circuit analysis; we have to deal with the inductance, capacitance and resistance of the metal waveguides that make up the optical waveguide/metal waveguide structure. There are many beautifully engineered solutions to this problem that minimise the power required to produce the electric field change and maximise the bandwidth (minimise the time taken) [6]. However, a very significant portion of the energy consumed in sending data around the internet is still used in the process of moving electrons around in the metal waveguides sitting on top of optical wave-guides, as can be seen in the O/E/O overhead that is attached to CMOS switching in figure 1.1.

With photonic signals operating via the nonlinear optical effects to induce the refractive index changes, the photons are massless bosons and the inductance, capacitance and resistance concepts do not apply. With careful engineering it is possible to maximise the generally weak nonlinear effects yet large intensities implying large (very large compared to CMOS) switch energies, but the power involved need not be dissipated as heat and the time taken is only limited by the time required to move bound electrons, and that can be as short as a few attoseconds [7].

2.2 Electroabsorption and electrorefraction effects

The Franz–Keldysh effect refers to electrical field induced change in the optical absorption close to the band edge of a semiconductor. From a consideration [3, 8] of the electrical field induced overlap between the valence band and conduction band wavefunctions of a semiconductor, the following expression for the absorption coefficient spectrum with electric field, F, applied can be derived

$$\alpha(\omega, F) = \left(\frac{C_p \theta_F^{\frac{1}{2}}}{\omega}\right)\left[|Ai'(\eta)|^2 - \eta |Ai(\eta)|^2\right], \qquad (2.1)$$

where $\theta_F^3 = \frac{e^2 F^2}{2 m_{vc}\hbar}$, $\eta = \frac{E_g - \hbar\omega}{\hbar\theta_F}$, C_p is a material constant which is similar for III–V semiconductors and CGS units is given by $C_p \sim 1.76 \times 10^{12}$ cm^{-1}s$^{-1/2}$—actually, this value is for GaAs. m_{vc}, is the reduced mass of the electron–hole pair for GaAs, $m_{vc} = 0.065\, m_o$, where m_o is the mass of the free electron. ω is the optical angular frequency. Ai is the Airy function.

In addition, by making use of the Kramers–Kronig transformation, the change in refractive index induced by the electric field, F, the electrorefraction, can be obtained [2]

$$\Delta n(\omega, F) = n(\omega, F) - n(\omega, 0) = \frac{c}{\pi}\int_0^\infty \frac{\Delta\alpha(\omega, F)}{\Omega^2 - \omega^2}\,d\Omega. \qquad (2.2)$$

So all the regions of the spectrum in which the optical absorption is changed by the Franz–Keldysh effect contribute to a change in the refractive index.

The electroabsorption effect is the fundamental basis of electroabsorption modulators, (EAMs), which are extensively used for high speed digital encoding of light. So for our example that we will use to illustrate the Franz–Keldysh effect we take a quaternary semiconductor alloy, In Al$_x$Ga$_{1-x-y}$As, lattice matched to InP. This alloy can have its band-gap tuned to the wavelengths appropriate for optical fibre transmission (around 1550 nm). With the Al fraction of the InAl$_x$Ga$_{1-x-y}$As alloy set to $x = 0.05$ and then the band-gap is around 1500 nm and the absorption spectrum of this semiconductor is given by using equation (2.1). A plot of the absorption spectrum of this alloy is shown in figure 2.2 with no applied field and with an applied field of 10 kV cm^{-1}.

As can been seen from figure 2.2, the absorption for wavelengths close to the band edge of the semiconductor increases with applied field; this is the basis of the Franz–Kelydysh electroabsoption modulator (EAM). Also, from this figure a couple of points are apparent that should be borne in mind when designing an EAM. Even with the best of growth techniques and also because a reversed bias pn junction is usually employed to apply the electric field, there is a built-in electric field associated with the pn junction. So even before an external electronic signal is applied, the modulator does not start with the band edge looking as it does in the figure with $F = 0$. The band edge usually looks like it starts with some electric field already there. This consideration impacts on the electroabsorption figure of merit [9], which is the ratio of the applied electric field induced absorption to the existing absorption, $\frac{\Delta\alpha(\omega, F)}{\alpha(\omega, F)}$, where $\Delta\alpha(\omega, F) = \alpha(\omega,F_2) - \alpha(\omega,F_1)$, where F_2 is the applied field and F_1 is the built-in field in the pin junction.

Franz- Keldysh effect in InAlGaAs

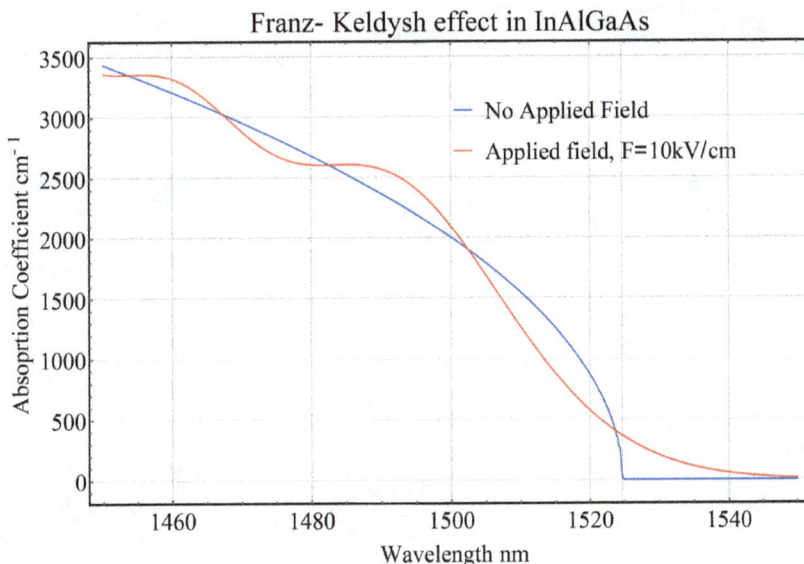

Figure 2.2. Comparison of the calculated absorption spectrum of InAlGaAs with Al fraction 0.05 with no applied electric field and with an applied field. See Mathematica notebook FK&FOM@telecoms wavelengths.cdf.

2.3 The Electroabsorption modulator (EAM)

The linear electro-optic effect and the electroabsoprtion effect have been utilised for optical communication systems usually employed in integrated optical devices. In this section we cover the electroabsorption modulator, a semiconductor integrated optics device.

2.3.1 The electroabsorption modulator (EAM) Franz–Keldysh effect in bulk semiconductor

Figure 2.3 illustrates the EAM waveguide device using the InAlAs alloy. The light is coupled into the input end of the waveguide from, for example, an optical fibre, or the EAM can form part of a semiconductor laser chip. The electrical signal that represents the data to be encoded onto the light is applied to the chip via an electrical input and on the mesa that also confines the light in the lateral direction. The design of the EAM is such that even modest voltages of say around 5 V, can result in large electrical fields across the reverse biased pin section of the EAM of around 50 kV cm^{-1}. For digital encoding of the light, the change in optical absorption is used to switch on and off the light transmitted by the EAM waveguide as in On Off Keying (OOK) [10].

A figure of merit for an EAM is the $\Delta\alpha/\alpha$ ratio, which, for an acceptable amount of modulation without excessive loss of optical signal, needs to be better than about 10 or so [9]. For efficient modulation it is also a good idea to try to minimise the electric field required to achieve the condition $\frac{\Delta\alpha}{\alpha} > 10$ (see figure 2.4).

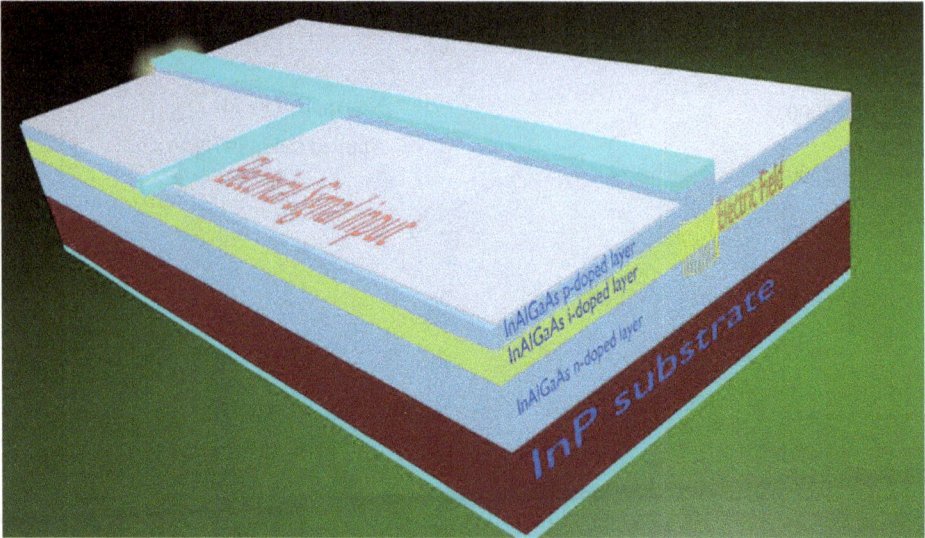

Figure 2.3. The layout of a typical EAM chip. The usual dimensions are that the chip is ~0.5 mm long, the optical waveguide strip on top of the chip is about 4 µm wide, The light is confined to the intrinsically doped (i-doped) region underneath the waveguide stripe. The electric field is applied underneath the waveguide strip across the InAlGaAs region that has its absorption spectrum altered by the Franz–Keldysh effect.

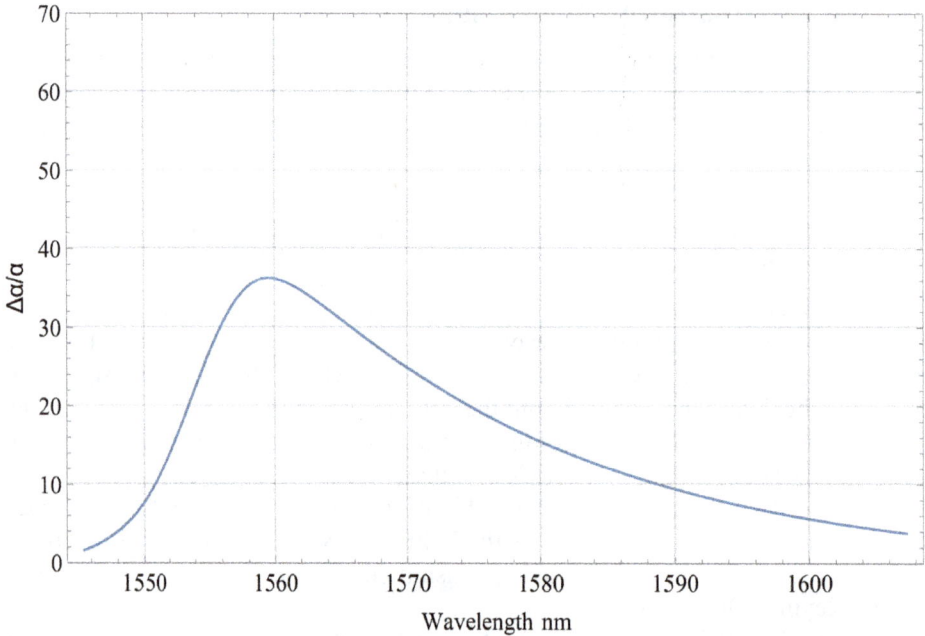

Figure 2.4. For the InAlGaAs alloy with Al fraction of 0.04, the modulation figure of merit for a background field of $F1 = 2$ Kv cm^{-1} and an applied field $F2 = 40$ Kv cm^{-1} the background loss is 8 cm^{-1} is plotted and for an EAM with low loss and good modulation depth $\Delta\alpha/\alpha > 10$, so for these conditions that is wavelengths 1550–1590 nm. See Mathematica notebook FK&FOM@telecoms wavelengths.cdf.

In fact, some the best materials for electroabsorption modulators are materials that make use of quantum confinement of electrons in semiconductor alloy hetero-structures [11]. The quantum confined structures evolved out of semiconductor heterostructures that were first employed in semiconductor lasers. If the small band-gap material in, for example, the heterostructure $Al_xGa_{1-x}As/Al_yGa_{1-y}As/Al_xGa_{1-x}As$ where $x>y$ has the small band-gap material layer $(Al_yGa_{1-y}A)$ is thinner than around 10 nm, then a quantum well is formed; the scattering of the electron wavefunction becomes less important and the wave nature of the electron requires that there is a resonant condition so that the wavefunction fits the quantum well. Along with the change in the physics there is a change in terminology and when dealing with the effect of electric fields on the optical absorption in quantum wells—it is referred to as the quantum confined Stark effect (QCSE). Compared to bulk semiconductors, quantum confined structures confer the ability to optimise the response of a material in a given wavelength region, see for example [12]. Quantum confined structures for linear optics, electroabsorption, electro-refractive and nonlinear optics are huge topics where this book will not venture—in part to keep the book concise and in part because the effects that are discussed in bulk semiconductors, although they may not be optimum, are more easily implemented since they do not require ultrasmall structures.

2.4 Electro-refractive modulators

The directional coupler [13] and the Mach–Zehnder interferometer [14] are both integrated optics devices that use LEO to modulate light. Both can be produced in AlGaAs waveguides and rely on LEO to electrically modulate the relative phase of light propagating in coupled waveguides. The devices convert an electrically induced phase change into an amplitude modulation. In the directional couplers the wave-guides are coupled because they are close enough that the modes in the waveguides overlap (see figure 2.5). In Mach–Zehnder interferometer the waveguides are coupled at Y junctions (see figure 2.6).

There has been a lot of interest in developing silicon as a photonic material so that the CMOS technology could be extended beyond electronics and achieve seamless integration with photonics [15]. In terms of electro-refractive effects, silicon, in its usual diamond-like crystal, is too symmetric to have an LEO effect and so only electrorefraction via changes in the absorption spectrum can be utilised [16]. The changes in the absorption spectrum are achieved either via the Franz–Keldysh effect or via free carrier induced absorption. The electrorefraction can then be estimated via the Kramers–Kronig transformation [17]. Usually in the form:-

$$\Delta n(V, F) = 6.3 \times 10^{-6}\,\text{cm} \cdot V \int_0^\infty \frac{\Delta\alpha(V', F)}{V'^2 - V^2}\mathrm{d}V',\qquad (2.3)$$

where V is in electron volts and $\Delta\alpha$ is in cm^{-1}—see also [2].

Although there has been significant progress with silicon photonic modulation devices, currently the optoelectronics is still largely dominated by direct-gap III–V semiconductors because they can be more easily integrated into direct-gap

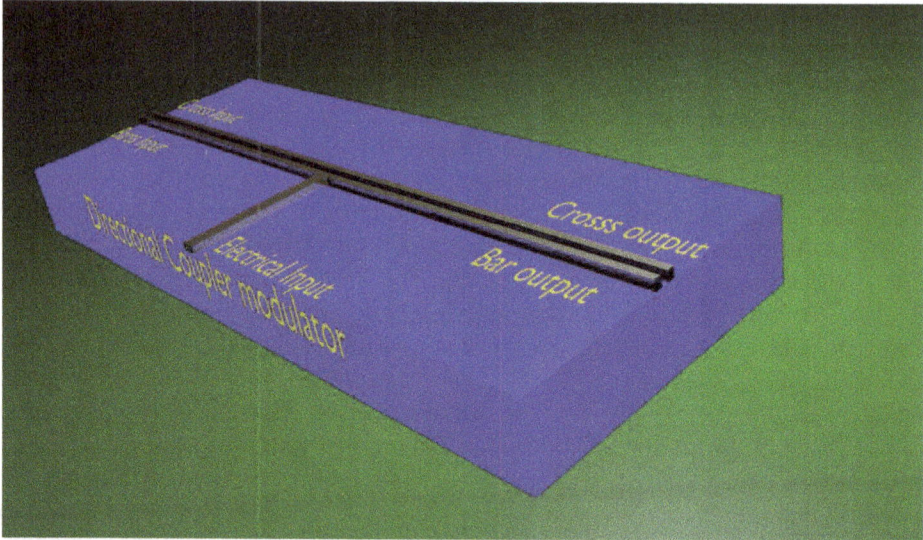

Figure 2.5. Directional coupler modulator. The bar and cross waveguides are in close enough proximity that their modes overlap. If the waveguides have the same propagation constant light input into the bar waveguide will couple over completely to the cross waveguide. An electrical signal input into the bar waveguide can change its propagation constant so that light no longer couples to the cross waveguide and thus the output light can be switched to the bar waveguide.

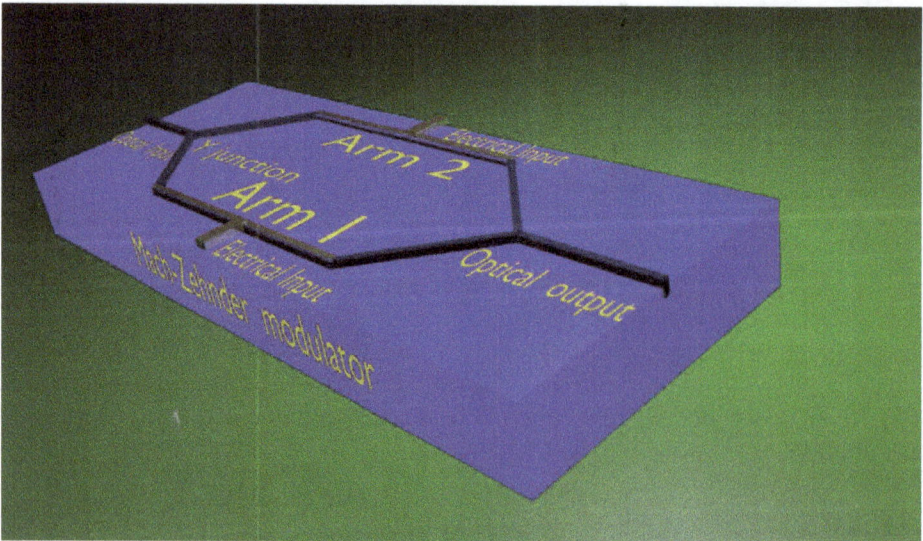

Figure 2.6. Mach–Zehnder interferometer modulator: the figure shows the waveguide layout for this device that consists of two back-to-back Y junctions with waveguides guiding light between the Y junctions. The first Y junction splits the input light into the arms of the interferometer and, via LEO, the electrical input signal controls the relative phase of the light at the second Y junction. The output light depends on the relative phase and the device can switch off the light by producing a destructive interference condition at the second Y junction.

semiconductor sources of light and, despite large investment, indirect-gap silicon remains relatively very inefficient as an emitter of light.

2.4.1 The directional coupler

The operation of the directional coupler modulator (illustrated in figure 2.5) depends on the overlap of the optical modes of two optical waveguides. If light is coupled into for example the bar waveguide it couples into the cross waveguide in an interaction length determined by the amount of mode overlap but this will only happen if the waveguides have the same propagation constant, $\beta = \frac{2\pi}{\lambda n_{\text{eff}}}$, where n_{eff} is the effective refractive index of the waveguide. With either LEO or electro-refraction it is possible to change the n_{eff} one waveguide so that they no longer have the same propagation constant. So an applied electric field to the bar waveguide can stop the input light from coupling to the cross waveguide and thus modulate the light.

For a semiconductor, the relative contribution of LEO or electrorefraction to the change in n_{eff} depends on how close the photon energy is to the band-gap—for photons energies just below the band-gap energy then electrorefraction dominates; for photon energies far from the band-gap energy LEO dominates [2]. The directional coupler has been demonstrated in AlGaAs using LEO [13].

A more detailed analysis of the directional coupler operation is given in the context of the all-optically switched nonlinear coupler in section 4.3.

2.4.2 The integrated electro-refractive Mach–Zehnder interferometer

The operation of the Mach–Zehnder interferometer modulator (illustrated in figure 2.6) depends on the phase of the light from each arm of the interferometer being controlled by the LEO. At the optical output Y junction the relative phase relationship between the light from each arm of the interferometer is converted into an amplitude modulation. With the LEO, if a voltage of opposite sign is applied to the arms of the interferometer then the induced change in n_{eff} has opposite sign and thus the phase in one arm is pushed in one direction and in the other arm it is pulled in the opposite direction.

The Mach–Zehnder interferometer modulator has been demonstrated in AlGaAs using LEO [14].

A more detailed analysis of the Mach–Zehnder interferometer operation is given in the context of the all-optically switched second order nonlinearity, push–pull Mach–Zehnder interferometer in section 4.2 and in section 4.4.

2.5 Conclusion

This chapter has covered LEO, electroabsorption and electrorefraction, and how these effects are employed in optical communication systems for modulation required for electrical–to–optical (E–O) data conversion used in optical communications and usually in OOK format.

LEO is closely related to second order nonlinear effects and electroabsorption and electrorefraction are closely related to third order nonlinear optics in as much as they are effects due to an applied electric field that changes the optical properties of the medium.

In the following chapters on all-optical switching we concentrate on optically induced phase changes via the cascade second order effect and the third order nonlinear refraction—sometimes called the optical Kerr effect. We will see how similar device configurations to the electro-refractive devices can be employed to convert optically induced phase changes into amplitude modulation or switching.

In this chapter we covered how electroabsorption and electrorefraction were related via the Kramers–Kronig transform. In chapter 3, we will also make use of the Kramers–Kronig relationship to show how absorptive optical nonlinearities associated with two-photon absorption are linked to the optical Kerr effect, a nonlinear refraction effect. This relationship turned out to be very important generally for optimising all-optical switching and we will take the example of AlGaAs optical waveguides to show in detail how it is used to optimise all-optical switching.

References

[1] Kerr J 1875 A new relationship between electricity and light: dielectrified media birefringent *Phil. Mag.* **50** 337

[2] Alping A and Coldren L A 1987 Electrorefraction in GaAs and InGaAsP and its application to phase modulators *J. Appl. Phys.* **61** 2430–3

[3] Tharmalingam K 1963 Optical absorption in the presence of a uniform field *Phys. Rev.* **130** 2204–6

[4] Li G L *et al* 1999 Ultrahigh-speed traveling-wave electroabsorption modulator—design and analysis *IEEE Trans. Microw. Theory Tech.* **47** 1177–83

[5] Miyamoto Y and Suzuki S 2010 Advanced optical modulation and multiplexing technologies for high-capacity OTN based on 100 Gb/s channel and beyond *IEEE Commun. Mag.* **48** 65–72

[6] Dogru S and Dagli N 2014 0.77-V drive voltage electro-optic modulator with bandwidth exceeding 67 GHz *Opt. Lett.* **39** 6074–7

[7] Hassan M T *et al* 2016 Optical attosecond pulses and tracking the nonlinear response of bound electrons *Nature* **530** 66

[8] Seraphin B O and Bottka N 1965 Franz–Keldysh effect of the refractive index in semiconductors *Phys. Rev.* **139** A560–5

[9] Chin M K 1992 On the figures of merit for electroabsorption wave-guide modulators *IEEE Photon. Technol. Lett.* **4** 726–8

[10] McMeekin S G *et al* 1994 Franz–Keldysh effect in an optical wave-guide containing a resonant-tunneling diode *Appl. Phys. Lett.* **65** 1076–8

[11] Chin M K 1995 Comparative-analysis of the performance limits of Franz–Keldysh effect and quantum-confined Stark-effect electroabsorption wave-guide modulators *IEE Proc.-Optoelectron.* **142** 109–14

[12] Kuzyk M G, Perez-Moreno J and Shafei S 2013 Sum rules and scaling in nonlinear optics *Phys. Lett.* **529** 297–398

[13] Campbell J C *et al* 1975 GaAs electro-optic directional-coupler switch *Appl. Phys. Lett.* **27** 202–5

[14] Buchmann P *et al* 1985 Broad-band Y-branch electro-optic GaAs wave-guide interferometer for 1.3 microns *Appl. Phys. Lett.* **46** 462–4

[15] Reed G T *et al* 2010 Silicon optical modulators *Nat. Photon.* **4** 518–26

[16] Nedeljkovic M, Soref R and Mashanovich G Z 2015 Predictions of free-carrier electro-absorption and electrorefraction in germanium *IEEE Photon. J* **7** 14

[17] Soref R A and Bennett B R 1987 Electro-optical effects in silicon *IEEE J. Quantum Electro.* **23** 123–9

Chapter 3

Nonlinear refraction

3.1 Introduction

When an electric field is applied to a material it produces a polarisation. The polarisation concept can be (here, we explicitly include the frequency dependence) expressed as:

$$p = \chi^{(1)}(\omega)E + \chi^{(2)}(\omega_1, \omega_2, \omega_3)EE + \chi^{(3)}(\omega_1, \omega_2, \omega_3, \omega_4)EEE... \quad (3.1)$$

This allows the nonlinear effects to be neatly incorporated into Maxwell's equations but it does not give insight into the mechanisms that give rise to $\chi^{(2)}(\omega_1, \omega_2, \omega_3)$ and $\chi^{(3)}(\omega_1, \omega_2, \omega_3, \omega_4)$. But the physical mechanisms will impact on how fast we can use nonlinear optics to switch light. There are a whole variety of these mechanisms from the slow thermal effects (~1 ms)—(thermal effects are limited by the thermal diffusion coefficient but small volumes promise a reduction in switching speed [1]) to ultra-fast effects due to the re-arrangements of the valence electrons [2], plus intermediate speed effects due to the rearrangement of free charge carriers as in silicon photonics [3]. In this book we concentrate on ultrafast mechanisms due to valence electrons. For a more comprehensive recent review, see [4].

We are particular interested in nonlinear refraction or more specifically intensity dependent phase changes in light propagation in guided wave formats. The nonlinear phase effect without concomitant linear or nonlinear absorption offers the exciting possibility of ultrafast switching without heating because, as we have seen with CMOS technology, it is heat that kills speed of processing.

Second order nonlinearity, $\chi^{(2)}(\omega_1, \omega_2, \omega_3)$, is traditionally associated with three-wave mixing to produce new wavelengths, but there are a variety of ways the second order nonlinearity can be employed for all-optical signal processing [5]. Below we discuss how it can be employed to give a nonlinear phase effect for all-optical switching.

The third order nonlinear effect, $\chi^{(3)}(\omega_1, \omega_2, \omega_3, \omega_4)$, is associated with the nonlinear refractive index or optical Kerr effect. In this chapter we will see how this is connected in semiconductors to two-photon absorption and how we can optimise the nonlinear refraction so that it is the pure optical Kerr effect with minimum nonlinear absorption. Further, the two-photon absorption process can be run backwards to give two-photon gain and this could lead to an important technology for generating the type of ultrashort high intensity pulses required for high speed, all-optical processing.

3.2 Cascade second order optical nonlinearity nonlinear phase change

3.2.1 Theory of cascade second order optical nonlinearity

The cascade second order effect as a way of producing a nonlinear phase change has shown promise for several applications, including all-optical switching, and much of the early work was carried out by Professor Stegeman's group at the University of Central Florida [6].

In a 1962 classic paper [7], Armstrong *et al* describe the formulism for three-wave mixing for second order nonlinear optical effects. In the common second harmonic generation configuration this paper describes how the energy exchanged between the fundamental and the second harmonic depends on various material parameters and the intensity of the fundamental. However, the paper misses out on how, if the phase matching condition is not completely satisfied, the phase of the fundamental can be altered when the energy of the fundamental is converted to the second harmonic and then converted back again. This effect is called the cascade second order nonlinear effect because the fundamental is converted to the second harmonic, propagates in the nonlinear crystal for a distance determined by the intensity of the fundamental and the magnitude of the phase mismatch, and then is converted in the reverse process back to the fundamental. So energy exchanges (or cascades) between the fundamental and the second harmonic as the light propagates in the second order crystal. Using a similar formalism to Armstrong *et al* [7], Ironside *et al* [8] described how the phase of the fundamental is altered when the phase matching condition is not completely satisfied.

In the analysis of this problem, we are dealing with nonlinear coupled differential equations and we will encounter the elliptic functions that we will encounter again with the analysis of the nonlinear coupler in chapter 4, where again we deal with nonlinear coupled differential equations. For a general discussion on elliptic functions, see [9].

Here we give a summary of [8]. In the second order, $\chi^{(2)}$ three-mixing process where light at the fundamental frequency ω is frequency doubled to 2ω, we can in general expect a phase mismatch between the fundamental wave and the second harmonic as they both propagate. The phase mismatch, Δk, can be expressed in terms of the propagation constants as follows:-

$$\Delta k = k_{2\omega} - 2k_{\omega}, \tag{3.2}$$

where $k_\omega = \frac{2\pi n_\omega}{\lambda}$, $k_{2\omega} = \frac{2\pi n_{2\omega}}{\lambda}$ and we can expect that, in general, $n_\omega \neq n_{2\omega}$. For efficient second harmonic generation there are various techniques for achieiving the phase matched condition $\Delta k = 0$. These techniques include using a Bragg diffraction grating [10]—so called quasi phase matching—the Bragg grating has period Λ and the wavevector associated with the grating $K = \frac{2\pi}{\Lambda}$ then $\Delta k = k_{2\omega} - 2k_\omega - K$ and if $K = k_{2\omega} - 2k_\omega$ then $\Delta k = 0$ and that ensures efficient generation of the second harmonic. However, in the cascade effect the objective is not efficient second harmonic generation but an intensity dependent phase change of the fundamental, and that requires that $K \neq k_{2\omega} - 2k_\omega$.

In what follows in the cascade effect we operate with $\Delta k \neq 0$ we get a periodic exchange of energy between the fundamental and the second harmonic and the phase of the fundamental becomes intensity dependent. In the analysis of this effect we start with coupled differential equations. As was shown by Armstrong et al [7], the fundamental wave electric field, E_1, and the second harmonic wave electric field, E_2, are coupled through the second order nonlinearity $\chi^{(2)}$ as follows:

$$\frac{dE_1}{dz} = i\frac{k\chi^{(2)}E_2E_1^*}{2n_\omega}e^{i\Delta k_z} \tag{3.3}$$

$$\frac{dE_2}{dz} = i\frac{k\chi^{(2)}E_1^2}{2n_{2\omega}}e^{-i\Delta k_z}, \tag{3.4}$$

where z is the propagation direction and k is the free space propagation constant. The second harmonic field can be eliminated to give:

$$\frac{d^2E_2}{dz^2} = i\Delta k_z\frac{dE_1}{dz} - \Gamma^2\left(1 - \left|\frac{E_1^2}{E_0^2}\right|\right)E_1, \tag{3.5}$$

where

$$\Gamma = \frac{kdE_0}{\sqrt{(n_\omega n_{2\omega})}}, \tag{3.6}$$

where $d = \chi^{(2)}/2$ and E_0 is the fundamental electric field at the start of the process. As is shown in Ironside et al [8], the nonlinear phase induced by the second order nonlinearity in the fundamental is given by

$$\theta = -\frac{1}{\alpha}\Pi(D,\phi|m), \tag{3.7}$$

where Π is the elliptic function of the third kind and the various parameters in equation (3.17) [8] are defined as follows

$$\alpha = \frac{1 + 2\epsilon + (1 + 4\epsilon)^{1/2}}{\sqrt{2}} \tag{3.8}$$

$$\phi = \arcsin{(sn(\acute{Z},m))} \tag{3.9}$$

$$m = \frac{1 + 2\epsilon - (1 + 4\epsilon)^{1/2}}{1 + 2\epsilon + (1 + 4\epsilon)^{1/2}} \tag{3.10}$$

and $\epsilon = (2\Gamma/\Delta k)^2$, D is the depletion parameter of the fundamental, sn is a Jacobi elliptic function (see [8] for the definition of $\acute{Z}$).

z_p is the distance of the periodic exchange of energy between the fundamental and the second harmonic as they propagate in the medium with second harmonic nonlinearity—see figure 3.1—and is given by

$$z_p = \frac{4\sqrt{2}}{\Delta k}\left(1 + 2\epsilon + (1 + 4\epsilon)^{\frac{1}{2}}\right)^{\frac{1}{2}} K(m), \tag{3.11}$$

where $K(m)$ is the complete elliptic integral of the first kind.

At the end of the waveguide of length, L, the nonlinear phase shift, including the total phase shift, is given by

$$\theta_{NL} = \theta + \frac{\Delta k L}{2}. \tag{3.12}$$

3.2.2 AlGaAs waveguide example of the cascade second harmonic effect

As an example of how this effect results in a nonlinear phase change we take the integrated optical waveguide example in AlGaAs with a phase matching grating.

Figure 3.1 shows the layout of an integrated chip with an optical waveguide that incorporates a grating for phase matching. The period of the grating determines the phase match condition $\Delta k = k_{2\omega} - 2k_\omega - K$. For example, if we take parameters appropriate for an AlGaAs waveguide and the length of the waveguide $L = 10$ mm and we take the free space fundamental wavelength as 1550 nm then $k_\omega = \frac{2\pi n_\omega}{\lambda}$ with $n_\omega = 3.45$ and $2k_{2\omega} = \frac{2\pi n_{2\omega}}{0.5\lambda}$ with $n_{2\omega} = 3.6$ then $K = \frac{2\pi}{\Lambda}$, where Λ is the periodicity of the grating. We set the periodicity to determine the phase matching condition so for example to obtain a phase match condition $\Delta k L = 2\pi$ we need $\Lambda \sim 5$ µm.

As can been see from figure 3.2, the exchange length between the fundamental and second harmonic wave is intensity dependent, as is the depletion of the fundamental and the nonlinear phase change to the fundamental at the output from the waveguide.

The nonlinear phase shift means that by making use of the cascaded second order nonlinearity it is possible to mimic the third order phase change, but as we shall see, the cascaded second order effect requires much less intensity to produce the same nonlinear phase shift as the third order effect. It should be noted from figure 3.2 that the nonlinear phase change is not quite like the third order effect in that the change in phase is not entirely linear with intensity and can become sublinear, as was

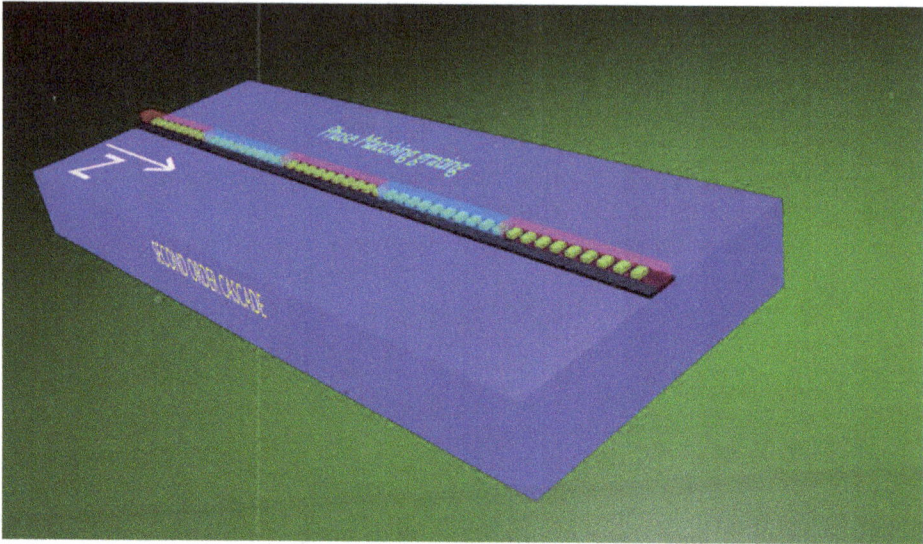

Figure 3.1. Schematic of second order cascade process in an optical waveguide with a phase matching grating. If the phase matching is nonzero, the fundamental (red) is converted to the second harmonic (blue) but converts back again, but the phase on the fundamental is altered by the process. Because the conversion between the fundamental and second harmonic is intensity dependent, the phase change is also intensity dependent.

noted in early experiments on the cascade effect where the phase change via a refractive index change was measured using the Z scan technique [11].

As can be seen from figure 3.3, the nonlinear phase change has the opposite sign from the phase mismatch, ΔkL. Therefore, the sign of the nonlinear phase change can be adjusted by altering the phase match condition. This feature will be used in the push–pull switch described in section 4.2.

An ultrafast all-optical device based on the cascade second effect has been reported [12, 13] using lithium niobate as the second order, $\chi^{(2)}$, medium and periodic poling for phase matching control.

3.3 Two-photon effects

The ultrafast, $\chi^{(3)}$, related mechanisms discussed here, are largely caused by multi-photon processes. It was a concept fundamental to the early development of quantum mechanics that matter and radiation could only exchange energy in discrete bundles of energy, photons, and the implicit assumption was that the exchange of energy would happen one photon at a time. However, some of the early pioneers [14, 15] of quantum theory were aware that energy could be exchanged in multi-bundles of energy and that two, three, etc. photon processes were possible. The probability of these multi-photon interactions increased with intensity and only became readily observable with the invention of the laser and the easy availability of high intensity light. One of the first multi-photon processes

(a) Exchange Length

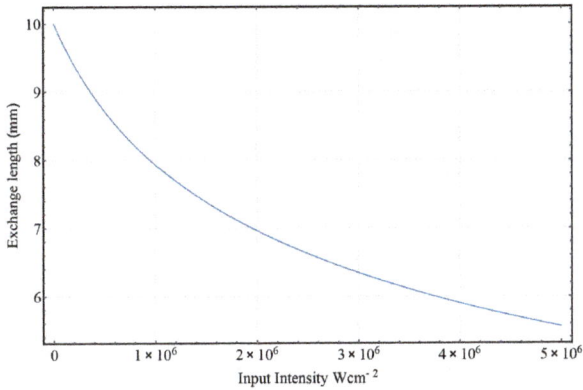

(b) Fundamental Depletion

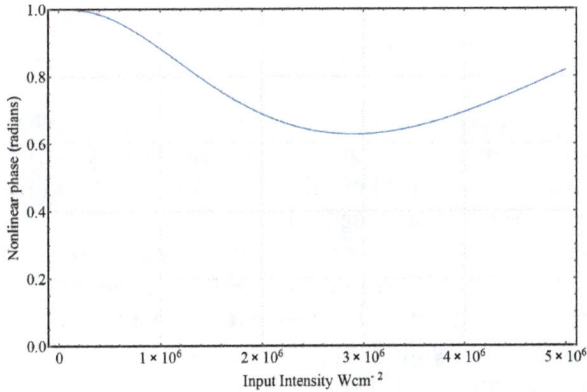

(c) Cascade Nonlinear Phase shift

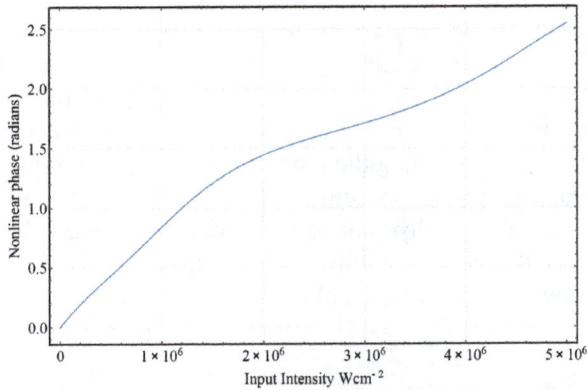

Figure 3.2. (a) Shows the exchange length, (b) the depletion of the fundamental and (c) the nonlinear phase change as a function of input intensity. These calculations are for an AlGaAs waveguide with the following parameters: fundamental wavelength 1550 nm, $n_\omega = 3.45$, $n_{2\omega} = 3.6$, second harmonic generation coefficient, $d_{14} = 150$ pmV^{-1}, length of waveguide, $l = 10$ mm, phase mismatch, $\Delta kl = -2\pi$. See notebook Cascaded 2nd order.nb.

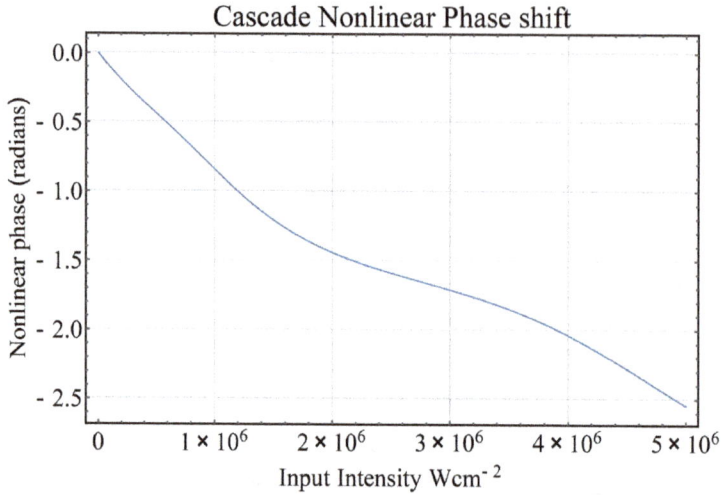

Figure 3.3. The cascade phase change identical conditions to figure 3.2 but with the phase mismatch of the opposite sign $\Delta kl = 2\pi$. See chapter 6 notebook Phase Missmatch.cdf.

to be observed was two-photon absorption; subsequently, two-photon effects have been developed for a host of photonic applications, including ultrafast pulse measurement; for a recent review see [16].

In semiconductors there are not only two-photon absorption effects, but in conditions far from thermal equilibrium where there is population inversion, two-photon absorption can operate in reverse and there is two-photon gain. A prediction of two-photon gain [17] in semiconductors has recently received experimental confirmation [18, 19].

In this section we will cover two-photon effects in semiconductors, how two-photon absorption relates to the optical Kerr effect, and we also cover two-photon gain in semiconductors. Since their invention, semiconductor lasers (and indeed lasers in general) have been stuck at $n = 1$ where n is the number of photons produced per transition, but early pioneers of lasers hoped and fully expected that lasers would move on to at least $n = 2$ [20]. This was because two-photon gain offers the exciting possibility of providing the ultimate in short pulse, high intensity sources of light with highly correlated photons. There has been some progress in atomic systems [21], but an electrically pumped two-photon semiconductor laser could provide photonics with a new, versatile, wavelength-agile source of highly coherent photons at high power in ultrashort pulses.

3.3.1 Two-photon absorption and nonlinear refraction

In semiconductors two-photon absorption has been extensively studied, [22, 23], and this has led to considerable insight into the process. Figure 3.4 gives a top level description of the process, in the two-photon absorption process two photons are

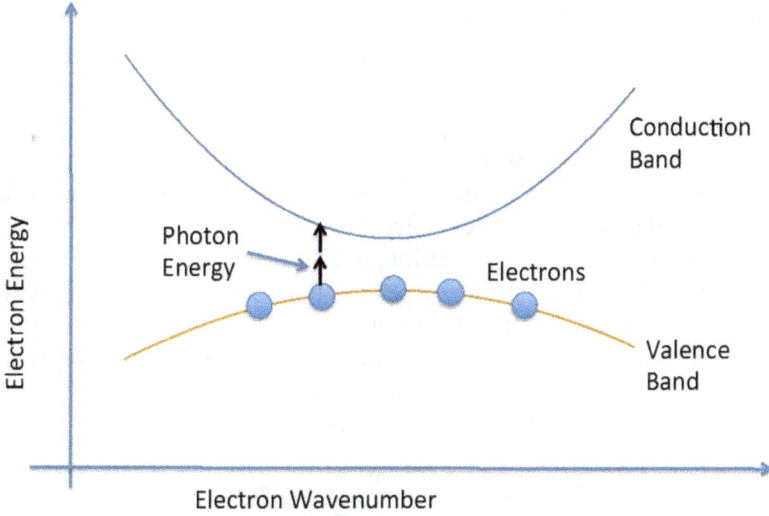

Figure 3.4. Schematic of two-photon absorption in a semiconductor. If the intensity is high enough two photons can be absorbed to span the band-gap energy of the semiconductor and promote an electron from the valence band to the conduction band.

absorbed to span the energy gap of the semiconductor. The process is intensity dependent and is only really observable with high intensity laser pulses.

Two-photon absorption is observed as an intensity dependent absorption according to the following equation

$$\frac{\mathrm{d}I}{\mathrm{d}z} = -\left(\alpha I + \beta_2 I^2\right), \tag{3.13}$$

where, I is the intensity of light, z, is the propagation direction, α is the linear absorption coefficient, β_2 is the two-photon absorption coefficient, which is given by

$$\beta_2 = \frac{2\hbar\omega_p}{I^2} M_{12}, \tag{3.14}$$

where $\hbar\omega_p$ is the photon energy, M_{12} is the transition rate for electrons from the valence band to the conduction band via the two-photon absorption process. The transition rate is worked out using second order perturbation theory [24] and the dispersion of the two-photon absorption coefficient was worked out to be given by:

$$\beta_2 = 3.1 \times 10^3 \frac{E_p^{1/2}}{n_0^2 E_g^3} F\left(\frac{2\hbar\omega_p}{E_g}\right), \tag{3.15}$$

where E_p is the Kane momentum energy—see [25]—E_g is the band-gap energy, n is the refractive index and the function $F(y)$ takes care of the dispersion of the two-photon absorption. If E_p and E_g are input in electron volts (eV) the factor 3.1×10^3 converts the answer to β_2 in cm GW^{-1}.

For most semiconductors of interest $F(y)$ is given by:

$$F(y) = \frac{(y-1)^{3/2}}{y^5}. \tag{3.16}$$

This theory for two-photon absorption has been largely successful in predicting two-photon absorption in a wide range of semiconductors. Furthermore, it has been expanded to include the optical Kerr effect [26].

The equation for nonlinear refraction, n_2, has been derived by applying the Kramers–Kronig relationship to the dispersion equation β_2 in a similar way to that discussed for linear optics, electroabsorption and electrorefraction. So from the two-photon induced change in the absorption, $\Delta\alpha$, it is possible to predict the two-photon induced change in the refractive index, $\Delta n = n_2 I$.

Explicitly, the equation for this is as follows:

$$\Delta n(\omega, \xi) = \int_0^\infty \frac{\Delta\alpha(\acute{\omega}, \xi)}{\omega'^2 - \omega^2} d\acute{\omega}, \tag{3.17}$$

where ξ, represents whatever causes the change in the optical property; so for electrorefraction and electroabsorption that would be the applied electric field, F, and for the two-photon effects that would be the optical intensity. From this Kramers–Kronig relationship the dispersion of the intensity dependent refractive index, n_2, has been determined to be [26]:

$$n_2 = 62.8 \frac{E_p^{1/2}}{n_0 E_g^4} G_2\left(\frac{\hbar\omega_p}{E_g}\right) \tag{3.18}$$

If E_p and E_g are input in electron volts (eV) the factor 62.8 converts the answer to n_2 in cm^2 W^{-1}.

The function, $G_2(y)$, is given by

$$G_2(y) = \frac{-2 + 6y - 3y^2 - \frac{3}{4}y^4 - \frac{3}{4}y^5 + 2(1-2y)^{3/2}u(1-2y)}{n_0 E_g^4} \tag{3.19}$$

This function determines the dispersion of n_2 ($u(x)$ is the unit step function).

Although highly successful and remarkable in its range of applications (see figure 3.5) the limits of this theory should be borne in mind. It was developed from a consideration of zinc blend direct gap semiconductors and mechanisms that involve bound electrons so mechanisms where free electrons are important are not covered by this theory.

Even the nonlinear optical response of bound electrons is not entirely instantaneous because if it was instantaneous there would be no dispersion, as described by equation (3.29). So clearly the response is not instantaneous and it has recently been measured [2]—it is 115 attoseconds (115 × 10^{-15}s). From the point of view of ultrafast all-optical information processing technology, that would appear to be the speed limit for data switching.

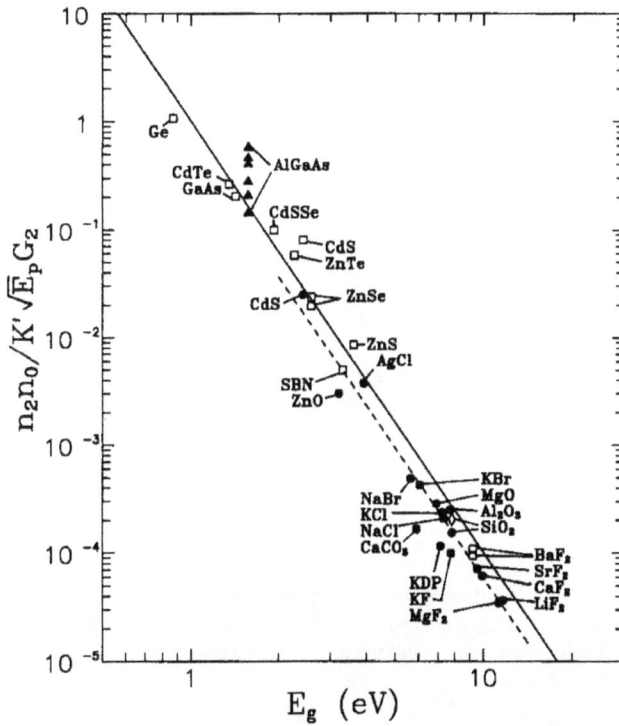

Figure 3.5. A log–log plot from that shows for a wide range of materials the expected E_g^{-4} dependence (see equation (3.29)) of a normalised n_2. © 1991 IEEE. Reprinted, with permission, from [26].

3.3.2 The $Al_xGa_{1-x}As$ example

Using the above theory, that is, by using the equations for β_2 (3.15) and n_2 (3.18), we can select the best semiconductor alloy, or more precisely a band-gap energy for a given wavelength region, that is, we can select an alloy that minimises the two-photon absorption, β_2, while maintaining n_2.

The theory of two-photon absorption and the nonlinear refractive index can be very nicely illustrated if we take the semiconductor alloy $Al_xGa_{1-x}As$ as an example. We take the wavelengths of interest as the optical communications wavelengths, which are determined by the optical properties of optical fibre, the minimum absorption and dispersion characteristics; that turns out to be the range 1260–1675 nm. The most commonly used wavelengths, the so-called C band, is the range 1530–1565 nm. Also, we keep the Al fraction, x, in the range 0–0.44. If $x > 0.44$, the $Al_xGa_{1-x}As$ alloy has an indirect band gap and that is outside the range of validity of our theory as presently formulated.

By adjusting the Al fraction it is possible optimise the nonlinear optical properties of the $Al_xGa_{1-x}As$ alloy for optical communications wavelength (around 1550 nm), as is illustrated in figure 3.6.

Ideally, for an ultrafast all-optical switch we require a large nonlinear refractive index free of any nonlinear absorption—more concisely stated, we need to maximise n_2 and minimise β_2, or maximise the $n_2 : \beta_2$, ratio. To avoid mixing up units we

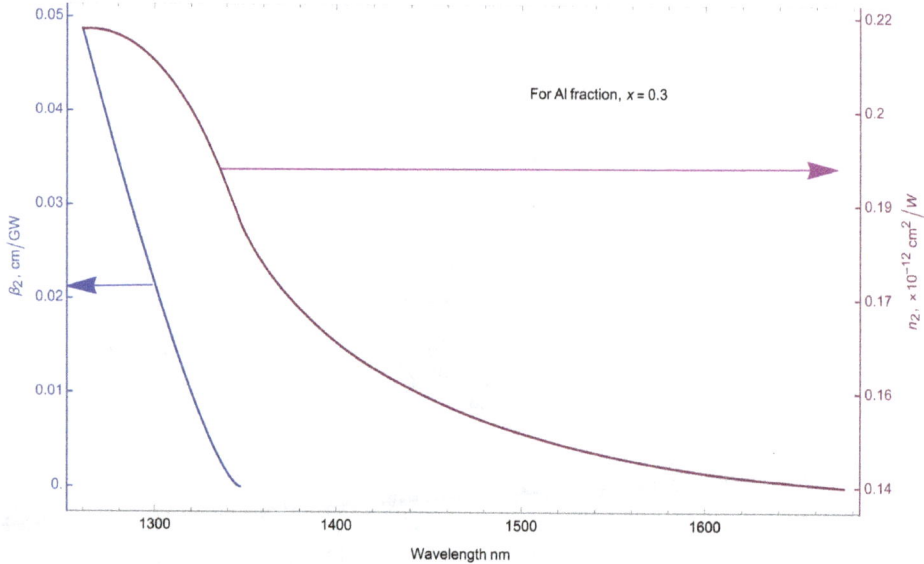

Figure 3.6. $Al_xGa_{1-x}As$ alloy: With $x = 0.3$ we show the two-photon absorption coefficient, β_2 and the nonlinear refractive index, n_2, versus wavelengths for optical communications. See the notebook Two Photon absorption and n$_2$ AlGaAs dispersion.nb and for a CDF of just the dispersion of n_2 as a function of x see Dispersion of n$_2$ for AlGaAs.cdf.

simply try to maximise the $G_2(y) : F(y)$ ratio for wavelengths around 1550 nm by adjusting the Al fraction, x, in the $Al_xGa_{1-x}As$ alloy. In figure 3.7, we have also included the fact that there will always be some linear loss present in the material—if only because of the scattering of light.

It turns out that for the $Al_xGa_{1-x}As$ alloy semiconductor the Al fraction, x, that optimises the optical properties for all-optical switching at around the 1450–1500 nm wavelengths is about $x = 0.15$ (see figure 3.7). Alloy fractions close to this value have been used in the optical devices that will be employed for all-optical ultrafast switching, and are described in sections 4.3 and 4.4.

Of course, the simple way to think about the figure of merit is that for a pure optical Kerr effect with no two-photon absorption then the operating photon energy should be less than half the band-gap energy. This rather begs the question, what about three-photon absorption? Indeed, these other multi-photon absorption processes can also limit all-optical switching devices but they are less significant than two-photon absorption [27].

3.3.3 Two-photon gain

Two-photon gain offers the possibility of providing the ultimate in short pulse high intensity sources of light with correlated photons. There has been some progress in atomic systems [21], but an electrically pumped two-photon semiconductor laser could provide photonics with a new, versatile, wavelength-agile source of highly coherent photons at high power in ultrashort pulses.

Switching figure of merit for $Al_xGa_{1-x}As$

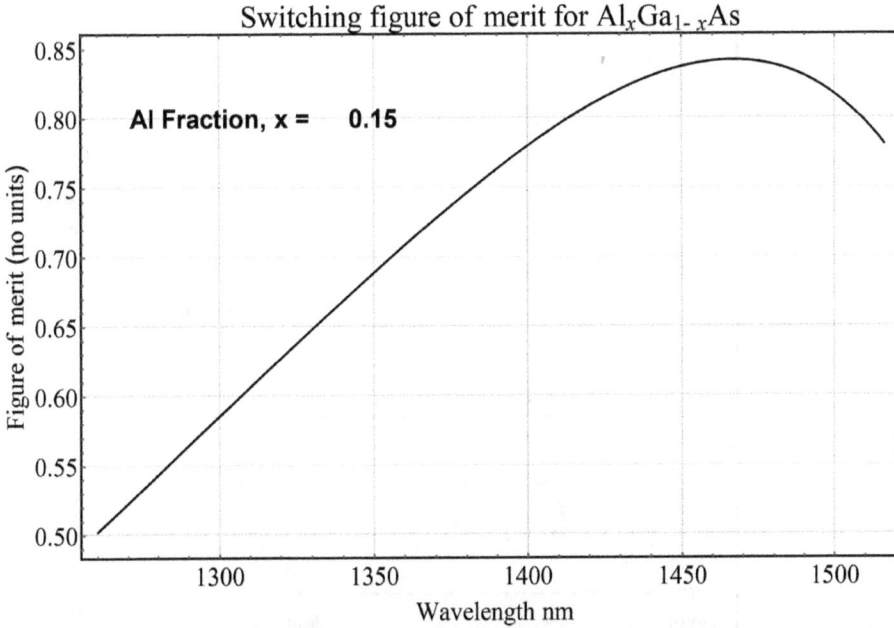

Figure 3.7. A plot of the switching figure of merit of $Al_xGa_{1-x}As$ material versus wavelength, for optical fibre communications wavelengths . See notebook Figure of Merit dispersion copy.nb.

As explained in [17], and as illustrated in figure 3.8, two-photon gain is just two-photon absorption running in reverse, and to reverse two-photon absorption requires that there is population inversion. For semiconductors we can take the two-photon absorption theory described in section 3.3 and combine it with the theory of semiconductor lasers where typically a quasi-Fermi level approach is used to determine population inversion and then we can estimate the size of the effect plus come up with designs for two-photon amplifiers.

We will remain consistent with our AlGaAs theme and discuss a design of a two-photon amplifier using an AlGaAs operating at the telecommunications wavelengths and estimate its performance.

$$\gamma_2 = 3.1 \times 10^3 \frac{\sqrt{E_p}}{n^2 E_g^3} F\left(\frac{2\hbar\omega_p}{E_g}\right), \tag{3.20}$$

where F is the function given in equation (3.16) and if E_p and E_g are given in eV then the factor 3.1×10^3 converts the units of γ_2 to cm GW^{-1}. If we assume that we are dealing with degenerate two-photon gain, that is, both of the emitted photons have the same energy, then the expression for the maximum two-photon gain coefficient is

$$\gamma_2 = 3.1 \times 10^3 \frac{\sqrt{E_p}}{n^2 E_g^3} F\left(\frac{E_{FC} - E_{FV}}{E_g}\right), \tag{3.21}$$

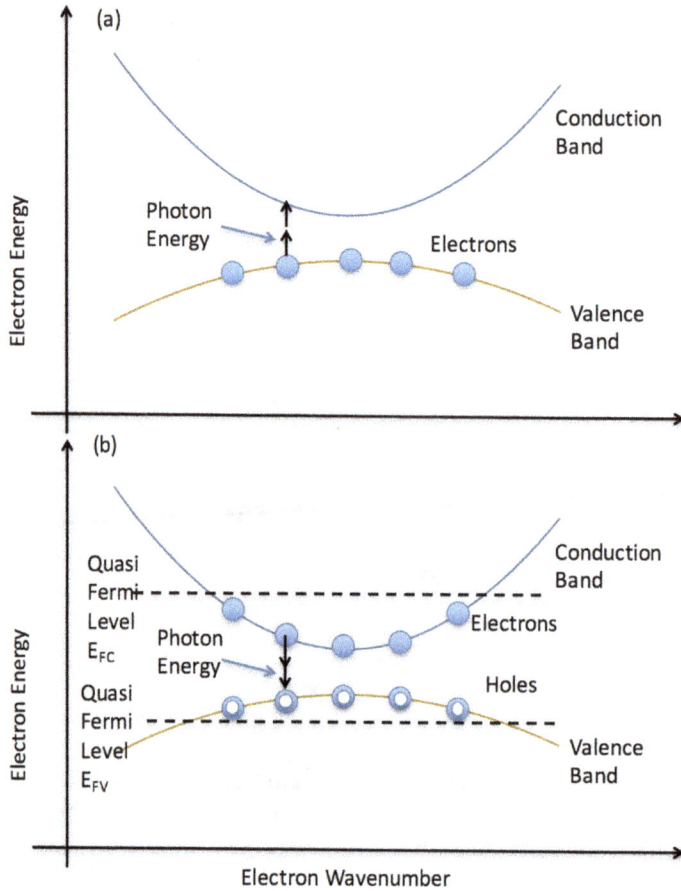

Figure 3.8. Schematic comparing (a) two-photon absorption with (b) two-photon gain. To bridge the band-gap energy two photons are required. For two-photon gain, population inversion in the semiconductor with electrical or optical pumping is required—the quasi-Fermi levels E_{FC} in the conduction, and E_{FV} in the valence band indicate the population inversion.

where E_{FC} and E_{FV} are the quasi-Fermi levels in the conduction and valence band, respectively. The quasi-Fermi levels can be predicted from using the following equation

$$E_{FC} = k_b T \left[\ln\left(\frac{n}{N_c}\right) + \sum_{i=1}^{\infty} A_i \left(\frac{n}{N_c}\right)^i \right], \tag{3.22}$$

where, n is the concentration of current injected electrons N_c is related to the density of states in the conduction band. A_i refers to the coefficients in the polynomial series expansion—for their values see [17]. There is a similar expression for E_{FV}.

With a combination of equations (3.21) and (3.22) it is possible to calculate the two-photon gain coefficient, as shown in figure 3.9.

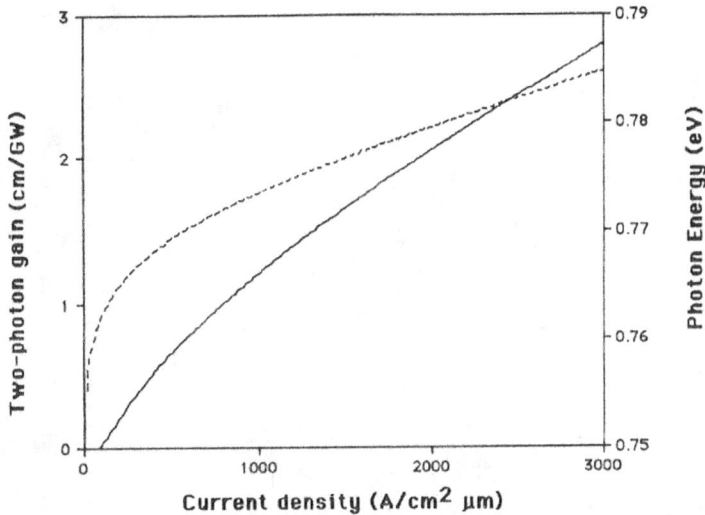

Figure 3.9. The two-photon gain characteristics for a GaAs double heterostruture at 77 K. The dotted line represents the photon energy that has maximum gain and the solid line represents the two-photon gain coefficient . © 1992 IEEE. Reprinted, with permission, from [17].

Experiments [19] on two-photon gain in an AlGaAs semiconductor waveguide that is electrically pumped to provide population inversion have confirmed the theory first described [17]. In the experiments an AlGaAs waveguide (see figure 3.10) was electrically pumped and the two-photon gain was measured by coupling in 100 fs pulses of light at a wavelength 1560 nm and observing the transmission as a function of the current injected into the waveguide. At current injection densities of 5000 Acm^{-2} µm the two-photon gain coefficient was measured as ~2.7 cm GW^{-1} in agreement with the predictions in [17].

However, as yet the exciting possibility of an oscillating two-photon semi-conductor laser has yet to be realised but it may be possible that by using so-called non-degenerate process then the two-photon threshold could be reached. Generally, in two-photon processes the two photons that are emitted (or absorbed) are not required to have the same energy—the so-called degeneracy condition. But the photons can have different energies (non-degenerate) as long as the sum of their energies is greater than the band-gap energy. For the correct arrangement of photon energies and intermediate states it has been shown [18] that it is possible to enhance the two-photon gain by orders of magnitude.

Because two-photon gain operates in reverse from two-photon absorption where high intensities are preferentially absorbed, two-photon gain preferentially amplifies high intensity pulses and does that on a very short time scale, again limited by the bound electron mechanisms, so around 115 × 10^{-15} s. Therefore, we can expect two-photon amplifiers to shorten even ultrashort pulses (<1 ps) and two-photon lasers to produce ultrashort pulses [28].

However, even in the non-degenerate case, it has not yet been possible to observe net two-photon gain in semiconductor waveguides. Primarily, due to competing

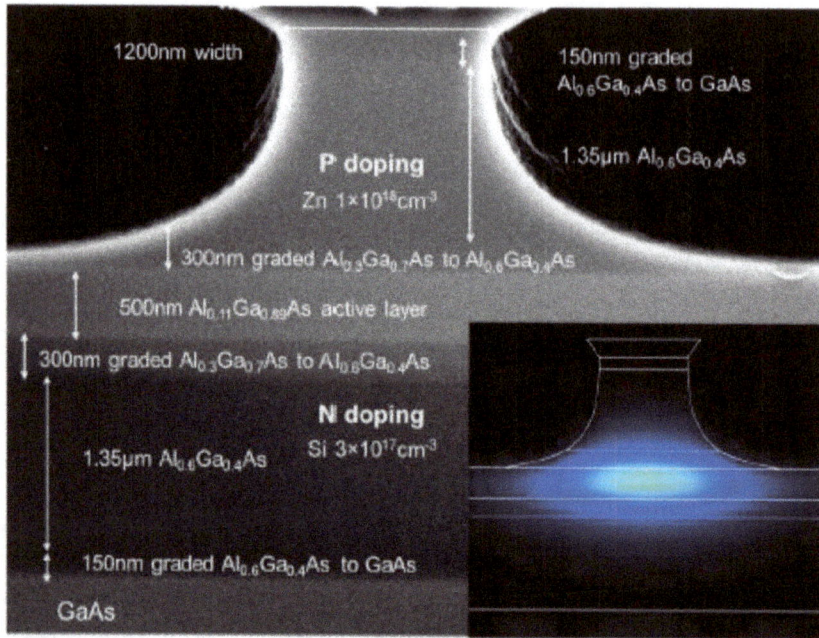

Figure 3.10. A labelled scanning electron micrograph (SEM) of an AlGaAs waveguide designed for measuring two-photon gain in an electrically pumped double heterostructure. The inset is a beam propagation simulation of the fundamental mode at a wavelength of 1560 nm. © 1991 IEEE. Reprinted, with permission, from [19].

nonlinear processes and probably free carrier absorption. So a two-photon semiconductor laser, and by that we mean a two-photon amplifier with sufficient gain to overcome losses and with sufficient feedback to oscillate, remains an unachieved challenge. Although it has been achieved in atomic systems [21].

3.4 Conclusions

In this chapter we have covered the basics of the nonlinear optical phenomena that need to be considered for ultrafast all-optical switching. The nonlinear optical effects we discussed were related to the second and third order optical nonlinearities, $\chi^{(2)}(\omega_1, \omega_2, \omega_3)$ and $\chi^{(3)}(\omega_1, \omega_2, \omega_3, \omega_4)$. The key consideration is to maximise the nonlinear refractive index while minimising the absorption, be that linear or nonlinear absorption.

In the cascade second order nonlinear optical effect the full analytical expression for the nonlinear phase shift was derived and an example based on an AlGaAs waveguide operating at telecommunications wavelengths was used to illustrate the derived equations. The absorption losses in the cascade nonlinearity are expected to be very small and essentially there is no heating of the device due to the cascade second order nonlinear optical effect.

The theory for the mechanisms behind the third order $\chi^{(3)}(\omega_1, \omega_2, \omega_3, \omega_4)$ optical nonlinearity was based on two-photon absorption in semiconductors. This theory uses the second order perturbation theory to derive the dispersion of the two-photon absorption coefficient, β_2, and then uses a nonlinear version of the Kramers–Kronig transformation to calculate the dispersion of nonlinear refraction, n_2. So for any given wavelength region, it becomes easy to work out the optimum band-gap that will minimise two-photon absorption coefficient, β_2, while maximising nonlinear refraction, n_2. We also saw that the theory, see figure 3.5, can be used to predict β_2 and n_2 for many materials. We illustrated the use of the theory with examples from the $Al_xGa_{1-x}As$ alloy.

From a fairly simple tweak to the theory that was derived for two-photon absorption in semiconductors it is possible to predict the two-photon gain characteristics of semiconductors, and recently this theory has been experimentally confirmed. According to this theory, there is huge potential for two-photon semiconductor lasers as sources of ultrashort high intensity of optical pulses. The challenge is in minimising competing linear and nonlinear absorption processes that have so far impeded the two-photon semiconductor laser from reaching laser threshold.

In the next chapter we show the second and third order optical nonlinearities, $\chi^{(2)}(\omega_1, \omega_2, \omega_3)$ and $\chi^{(3)}(\omega_1, \omega_2, \omega_3, \omega_4)$ can be employed in integrated optical formats to produce all-optical switching devices that can perform functions such as time division multiplexing on a picosecond pulse with minimum heating of the device because the nonlinear optical absorption is minimised. Further, within the limits of the theory presented for the dispersion two-photon absorption coefficient, β_2 and the nonlinear refraction, n_2 for optical communications wavelength it is certain that $Al_xGa_{1-x}As$ alloy is the best possible material, with the correct x fraction $x \sim 0.18$, for all-optical switching.

References

[1] Khurgin J B *et al* 2015 Ultrafast thermal nonlinearity *Sci. Rep.* **5** 8

[2] Hassan M T *et al* 2016 Optical attosecond pulses and tracking the nonlinear response of bound electrons *Nature* **530** 66

[3] Reed G T *et al* 2010 Silicon optical modulators *Nat. Photon.* **4** 518–26

[4] Wabnitz S and Eggleton B J 2015 *All-Optical Signal Processing* (Springer Series in Optical Sciences vol 194) (Berlin: Springer)

[5] Langrock C *et al* 2006 All-optical signal processing using $\chi^{(2)}$ nonlinearities in guided-wave devices *J. Lightwave Technol.* **24** 2579–92

[6] Stegeman G I, Hagan D J and Torner L 1996 $\chi^{(2)}$ cascading phenomena and their applications to all-optical signal processing, mode-locking, pulse compression and solitons *Opt. Quantum Electron.* **28** 1691–740

[7] Armstrong J A *et al* 1962 Interactions between light waves in a nonlinear dielectric *Phys. Rev.* **127** 1918–39

[8] Ironside C N, Aitchison J S and Arnold J M 1993 An all-optical switch employing the cascaded 2nd-order nonlinear effect *IEEE J. Quantum Electron.* **29** 2650–4

[9] Schwalm W A 2015 *Lectures on Selected Topics in Mathematical Physics: Elliptic Functions and Elliptic Integrals* (San Rafael, CA: Morgan & Claypool Publishers)

[10] Yoo S J B *et al* 1996 Wavelength conversion by difference frequency generation in AlGaAs waveguides with periodic domain inversion achieved by wafer bonding *Appl. Phys. Lett.* **68** 2609–11

[11] Desalvo R *et al* 1992 Self-focusing and self-defocusing by cascaded 2nd-order effects in KTP *Opt. Lett.* **17** 28–30

[12] Kanter G S *et al* 2001 Wavelength-selective pulsed all-optical switching based on cascaded second-order nonlinearity in a periodically poled lithium-niobate waveguide *IEEE Photon. Technol. Lett.* **13** 341–3

[13] Kanbara H *et al* 1999 All-optical switching based on cascading of second-order non-linearities in a periodically poled titanium-diffused lithium niobate waveguide *IEEE Photon. Technol. Lett.* **11** 328–30

[14] Dirac P A M 1927 The quantum theory of dispersion *Proc. R. Soc.* A **114** 710–28

[15] Göppert-Mayer M 1931 Über elementarakte mit zwei quantensprüngen *Ann. phys.* **401** 273–94

[16] Hayat A *et al* 2011 Applications of two-photon processes in semiconductor photonic devices: invited review *Semicond. Sci. Technol.* **26** 18

[17] Ironside C N 1992 2-Photon gain semiconductor amplifier *IEEE J. Quantum Electron.* **28** 842–7

[18] Reichert M *et al* 2016 Observation of nondegenerate two-photon gain in GaAs *Phys. Rev. Lett.* **117** 5

[19] Nevet A, Hayat A and Orenstein M 2010 Measurement of optical two-photon gain in electrically pumped AlGaAs at room temperature. *Phys. Rev. Lett.* **104** 4

[20] Prokhorov A M 1965 Quantum electronics *Science* **149** 828–30

[21] Gauthier D J 2003 Two-photon lasers *Progress in Optics* vol 45 ed E Wolf (Amsterdam: Elsevier) pp 205–72

[22] DeSalvo R *et al* 1996 Infrared to ultraviolet measurements of two-photon absorption and n_2 in wide bandgap solids *IEEE J. Quantum Electron.* **32** 1324–33

[23] Hutchings D C and Wherrett B S 1994 Theory of anisotropy of 2-photon absorption in zinc blende semiconductors *Phys. Rev.* B **49** 2418–26

[24] Pidgeon C R *et al* 1979 2-photon absorption in zincblende semiconductors *Phys. Rev. Lett.* **42** 1785–8

[25] Vanstryland E W *et al* 1985 2 photon-absorption, nonlinear refraction, and optical limiting in semiconductors *Opt. Eng.* **24** 613–23

[26] Sheikbahae M *et al* 1991 Dispersion of bound electronic nonlinear refraction in solids *IEEE J. Quantum Electron.* **27** 1296–309

[27] Kang J U *et al* 1994 Limitation due to 3-photon absorption on the useful spectral range for nonlinear optics in AlGaAs below half band-gap *Appl. Phys. Lett.* **65** 147–9

[28] Heatley D R, Firth W J and Ironside C N 1993 Ultrashort-pulse generation using 2-photon gain *Opt. Lett.* **18** 628–30

Chapter 4

Nonlinear optical devices

4.1 Introduction

As we have seen from the theory, significant nonlinear optical effects require high optical intensities and to maintain the high optical intensity over long interaction paths requires the light to be confined in optical waveguides. If the light is not confined it rapidly loses intensity as it diffracts. In addition, the optical waveguides should be as low loss as possible. So the simple physics of diffraction requires waveguides and these waveguides need to be engineered to minimise loss from absorption, linear and nonlinear absorption and scattering.

The obvious optical waveguide format to start with is the optical fibre; there has been a huge investment in this technology and linear losses can be as low as 0.2 dB km^{-1}. The band-gap of silica fibre is up around 9 eV, far enough away from the telecommunications wavelengths with photon energies around 0.7 eV to avoid any significant nonlinear absorption. So indeed the first of the guided wave, ultrafast all-optical switching devices were realised in optical fibre format [1]. The trade-off is that because there is such a big gap between the optical fibre band-gap and the photon energy, the size of the nonlinear refraction effect is small and therefore large interaction lengths are required to build up phase changes large enough for a switching operation. The magnitude of the nonlinear refractive index n_2 falls off as E_g^{-4}—see equation (3.19). So at 1550 nm, the silica based optical fibres have $n_2 \approx 10^{-19}$ cm^2 W^{-1} compared to, for example AlGaAs, with the band gap optimised at $n_2 \approx 10^{-14}$ cm^2 W^{-1}. The difference in n_2 comes more or less straight over into the scaling of the device, therefore, we can expect the AlGaAs device to be $\approx 10^{-5}$ of the length of the silica fibre devices.

There have been successful efforts to use optical fibre technology with larger n_2, by using chalcogenide glasses (the HNLF technology in figure 1.1) with a band-gap energy closer to the photon energy but far enough away to avoid two-photon absorption [2, 3]. The trade-off here is that the linear losses are not as low as silica

doi:10.1088/978-1-6817-4521-3ch4

fibre, but that is a limitation not imposed by the physics of nonlinear effects, it is more of a limitation imposed by the engineering of optical fibres from new materials.

There have been several reviews of ultrafast all-optical optical fibre devices [2, 4, 5]. In this book we are sticking with the theme of AlGaAs alloy and waveguide devices. In this chapter we will focus on AlGaAs semiconductor waveguide devices that employ both the second order and third order optical nonlinearities.

The second order device we discuss here employs the cascaded second order nonlinearity, as covered in section 3.2. The intensity dependent phase shift is given by equation (3.17). With III–V semiconductors in the crystalline zinc blend structure there is a second order optical nonlinearity than can be used for all-optical switching. The AlGaAs alloy does have a very significant second order nonlinear effect, $d_{14} \approx 150$ pm V^{-1} that has been employed in three-wave mixing devices. In this chapter we show how the AlGaAs alloy second order nonlinearity could be employed in an all-optical ultrafast switching device that uses the second order nonlinearity, the so-called cascade configuration, to produce an effective n_2 that can be much larger than the n_2 available from the third order optical nonlinearity.

For the third order nonlinear devices, the n_2 devices, many of the device concepts are essentially the same as for the fibre devices but the dimensions are much smaller, so it is just a matter of scaling down the fibre device to take advantage of the large n_2. In addition, the waveguides in the integrated devices need to be much smaller than the fibre devices because the linear losses, usually scattering losses, are generally much larger in semiconductor waveguides than in optical fibre. However, that is a limitation imposed by the engineering rather than the physics of the nonlinear effects.

4.2 The cascade second order optical switch—the push–pull switch

A version push–pull cascade second switch is shown in figure 4.1. where the chip layout of the device is illustrated. It shows an integrated Mach–Zehnder interferometer layout of optical waveguides on a chip. The interferometer consists of two back-to-back Y junctions separated by waveguides making up the arms of the interferometer. At the first Y junction the light is split evenly between the two arms and then the light propagates in the arms of the interferometer before it recombines at the second Y junction. The relative phase change of the light in each arm determines how much light is coupled into the last section of the device, the output waveguide. Each arm of the interferometer has a phase grating to control the phase mismatch between the fundamental and the second harmonic generated by the second order optical nonlinearity. In the push–pull configuration the gratings have different periods to produce a phase mismatch of equal magnitude but opposite sign in their respective arms. As we have seen in section 3.2.2, in the cascade second order nonlinearity, opposite phase mismatch gives an intensity dependent phase of opposite sign in each arm—thus push–pull in terms of the phase shift induced by high intensity. Figure 4.1 also shows high intensity control pulses (yellow) that alter the phase of the low intensity data pulses (red). The control pulses alter the phase of

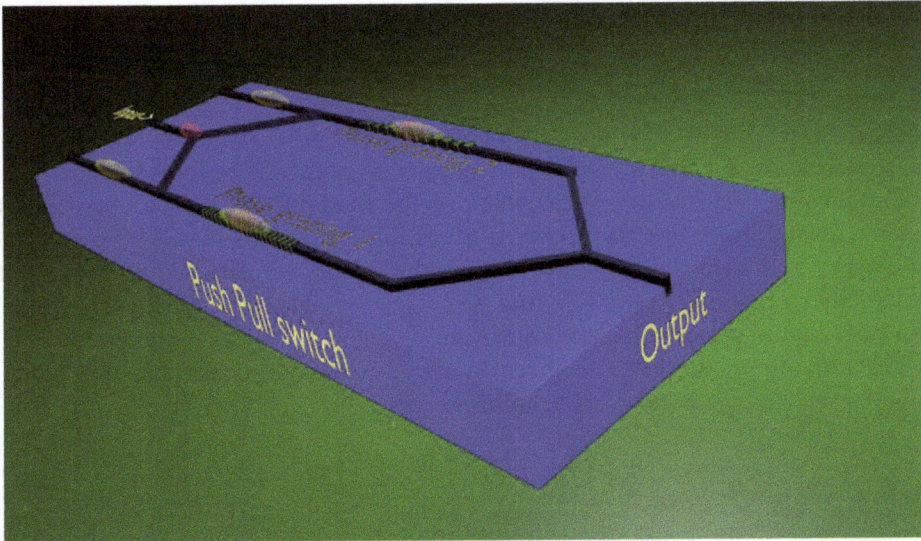

Figure 4.1. The integrated push–pull that uses the cascade second order nonlinear effect in a Mach–Zehnder interferometer configuration. The device uses gratings to control the phase match of the second order effect. The gratings have different periodicity and different phase matching characteristics with opposite signs. The control pulses are shown in yellow and the data pulses are shown in red.

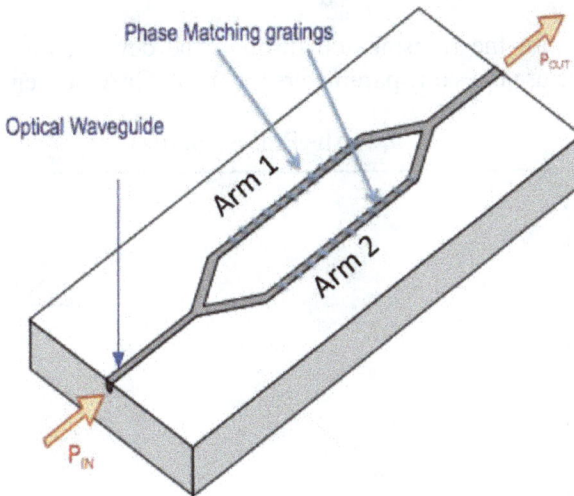

Figure 4.2. The simplified model of the push–pull switch without the control pulses.

the data pulses via the cascaded second order effect and thus can control the phase of the data pulses at the second Y junction where they recombine and switch between constructive and destructive interference.

As an example, we take the AlGaAs material discussed in section 3.2.2 and model the switching characteristics of the push–pull switch. We model a simplified version of the switch without the control pulses, as shown in figure 4.2.

The phase match grating controls the phase match, Δk as follows:

$$\Delta k = k_{2\omega} - 2k_{\omega} - K, \tag{4.1}$$

where $K = 2\frac{\pi}{\Lambda}$, Λ is the periodicity of the grating, $k_{2\omega}$ is the propagation constant of the second harmonic and k_{ω} is the propagation constant of the fundamental.

The intensity dependent optical phase of the fundamental that is obtained via cascade second order has been covered in section 3.2. In this push–pull switch the gratings in each arm (length, L) of the interferometer are designed with different periodicity so that $\Delta k L_{\text{arm1}} = -2\pi$ and $\Delta k L_{\text{arm2}} = 2\pi$. Thus the effective n_2 in each arm of the interferometer has opposite sign.

The response of the device is given by the Cos squared behaviour of a Mach–Zehnder interferometer adapted as follows:

$$T = (1 - D)\cos^2\left(\frac{\Delta\theta}{2} + \vartheta\right), \tag{4.2}$$

where D is the depletion of the fundamental, $\Delta\theta = \theta_{\text{NL1}} + \theta_{\text{NL2}}$ where θ_{NL1} is the phase change in arm 1 and θ_{NL2} is the phase change in arm 2. These nonlinear phases shifts are given by the following equation, similar to equation (3.7):

$$\theta_{\text{NL}} = -\frac{1}{\alpha}\Pi(D, \phi|m). \tag{4.3}$$

In figure 4.3 we plot the transmission through the device as a function of input intensity and if we use the same parameters for the AlGaAs waveguides discussed in

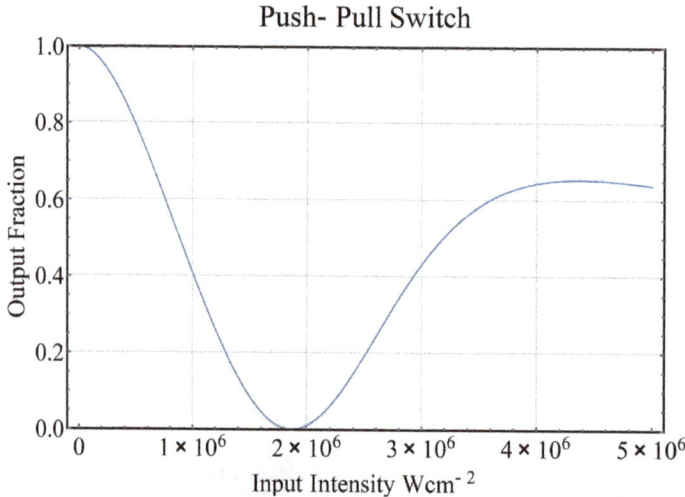

Figure 4.3. The graph shows a calculation of output fraction versus input intensity for the AlGaAs integrated push–pull switch based on [6]. The parameters used in the calculation operating fundamental wavelength 1500 nm, second order d coefficient, 150×10^{-12} mV^{-1}, refractive index of second harmonic wavelength (750 nm) $n_{2\omega} = 3.6$, refractive index of fundamental wavelength (1550 nm) $n_{\omega} = 3.45$, the phase mismatch $\Delta k L = \pm 2\pi$, the length of the arms $L = 0.01$ m. See Mathematica notebook Cascade Push-Pull Model.nb.

section 3.2.2 then we obtain a switch-off of the device when the input intensity is $\sim 2 \times 10^6$ W cm^{-2}.

The version of the push–pull switch modelled here has not been implemented, but similar devices in lithium niobate have shown promise and are switched with peak powers as low as 6.6 W in a 2.5 ps pulse, implying a switching energy of ~ 16 pJ [7].

4.2.1 AlGaAs Optical waveguides for n_2 switching

For all-optical switching that uses nonlinear refraction n_2, we use the $Al_xGa_{1-x}As$ alloy that is optimised for nonlinear refraction at wavelengths around 1550 nm. From the theory presented in section 3.3.2 the Al fraction with $x \sim 0.18$ gives a pure optical Kerr effect and minimises associated two-photon absorption. So the waveguides are designed to confine the high intensity light in $Al_{0.18}Ga_{0.82}As$ active layer over the length of the device and not limited by diffraction—the interaction length is in fact only limited by losses and the nonlinear losses are minimised so we are left only with linear absorption and scattering losses. The linear absorption is very low for photon energies just below half the band-gap energy. Scattering losses are governed by quality of growth and fabrication issues such as how smooth the walls of the waveguides are generally for the AlGaAs waveguide in the region of 1–10 cm^{-1}.

Figure 4.4 shows a schematic layout of a generic design of the AlGaAs waveguide for all-optical switching. The key point is that the active layer, the core layer of the waveguide, is the AlGaAs alloy optimised for a large nonlinear refractive index at ~ 1550 nm. The cladding layer is there to confine the light to the core and the buffer

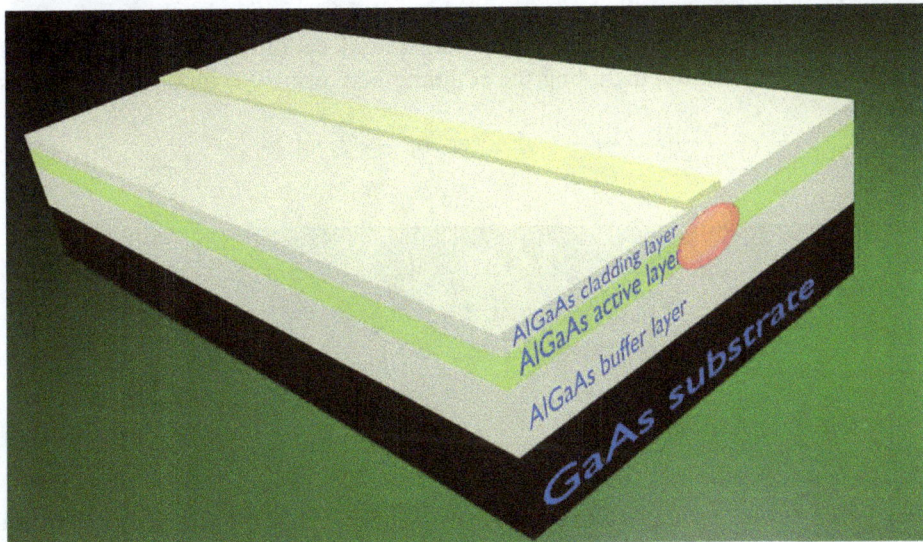

Figure 4.4. A schematic of the layout of an AlGaAs waveguide chip showing the layer structure and the mesa structure on top that confines the light laterally. The buffer layer is 4 μm thick $Al_{0.25}Ga_{0.75}As$. The active layer is 1.5 μm thick $Al_{0.18}Ga_{0.82}As$. The cladding layer is 1.5 μm thick $Al_{0.25}Ga_{0.75}As$. The mesa strip is 4 μm wide and 1.6 μm in height. The approximate position of the guided light at the facet is indicated by the red area.

layer is required to restrict the overlap of the active mode of the waveguide with the GaAs substrate [8].

4.3 The nonlinear directional coupler (NLC)

A simple layout of the AlGaAs waveguides can be used to switch light. Two waveguides in close enough proximity that their modes overlap will exchange light. For this exchange to take place the waveguides need to have the same effective refractive index. If a high intensity pulse is present in one of the waveguides then nonlinear refractive index effect results in the waveguides having different effective indices and the light no longer transfers between the waveguides. Figure 4.5 illustrates the concept and shows how the device could be used to switch data.

Time division multiplexing (TDM) is a common function required in many digital communication systems and is extensively employed in optical communications to route data. So for example in optical TCP-IP systems the information containing the address of the data is used to switch the data to the optical fibres that guide the data to its destination. In long haul fibre systems the data rates are in the range 10–40 GB s^{-1} and seem set to be continually increasing. Also, to access the enormous data carrying capability of optical fibre, TDM is often combined with wavelength division multiplexing (WDM). One of the attractions of all-optical switching is that it can use the attosecond bound electron nonlinearity [9] to achieve the ultimate in switching speeds and future proof the network against increases in data rates.

Here we present a summary of the theory first reported by Jenson [10] who used coupled mode theory to explain the operation of the nonlinear coupler.

The amplitudes of the modes in each guide are defined as a (for example, the amplitude of the mode in the bar waveguide) and $\acute{a}$ (for example, the amplitude of the mode in the cross waveguide); then the spatial evolution in the propagation direction, the z direction, including the coupling between these modes is as follows:

$$i\frac{\partial a}{\partial z} = Q_1 a + Q_2 \acute{a} + (Q_3 |a|^2 + Q_4 |\acute{a}|^2)a \qquad (4.4)$$

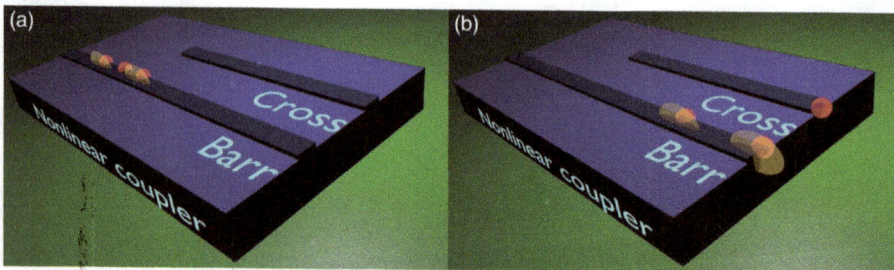

Figure 4.5. The figures illustrate the operation of the nonlinear coupler as time demultiplexer. (a) The red dots represent optical data pulses and yellow represent high intensity optical control pulses. The pulses are coupled into the bar waveguide. (b) The optical data pulses that have the control pulses in the same time slot stay in the bar waveguide. The other pulse moves to the cross waveguide. The high intensity pulses have changed the effective refractive index of the bar waveguide, detuned it from the cross waveguide and prevented the exchange of light. See also the animation available here.

$$i\frac{\partial \acute{a}}{\partial z} = Q_1\acute{a} + Q_2 a + (Q_3 |\acute{a}|^2 + Q_4 |a|^2)a. \tag{4.5}$$

The Q factors are various coupling factors as given by Jensen [10]. The coupling factors are essentially proportional to the overlap of the bar and cross waveguide modes—the further apart the guides, the smaller the coupling.

The appropriate boundary condition for a nonlinear coupler experiment is that all the input power is coupled into the bar guide, then, $I_b(z)$, intensity in the bar waveguide as a function of z, the propagation direction, is as follows:

$$I_b(z) = \frac{1}{2}I_b(z)\left[1 + cn\left(\frac{\pi z}{L_c}\bigg|m\right)\right], \tag{4.6}$$

where L_c is the linear coupling length that is the distance required for the intensity to couple over completely from the bar guide to the cross guide in the linear regime (low power input); $[cn(\frac{\pi z}{L_c}|m)]$ is the Jacobi elliptic function, $m = \frac{I_b(0)}{I_c}$, where, I_c, is the critical intensity defined as the intensity of the control pulse that will change the output from being completely in the cross waveguide to being equally in the bar and cross waveguides. For the nonlinear coupler the critical intensity is given by:

$$I_c = \frac{\lambda}{L_c n_2}. \tag{4.7}$$

The coupling length, L_c, is determined by the distance between the bar and cross waveguides that in turn determines the amount of overlap between the modes of the bar and cross waveguides. More overlap means shorter coupling length, L_c.

For ultrafast operation of the nonlinear coupler, the intensity dependent nonlinear refractive index, n_2, should be the entirely optical Kerr effect and therefore for operating at a wavelength around 1550 nm, the optimal material for the waveguide core is $Al_{0.18}Ga_{0.82}As$.

Operating at around 1550 nm, the first nonlinear coupler made from $Al_{0.18}Ga_{0.82}As$ as core was reported in 1991 [11]. The coupler was 6.25 mm long consisting of bar and cross waveguides of 4 µm width separated by 5 µm. The device was characterised using high intensity 10 ps pulses. The input intensity into the bar waveguide was varied up to maximum intensity of 2×10^9 W cm^{-2} and at the output of the coupler the intensity of light in the bar waveguide increased, and in the cross waveguide, the intensity decreased as the intensity at the input was increased. The nonlinear refractive index caused the coupling between the waveguides to diminish as the input intensity was increased.

The experimental results and the above theory (equation (4.6)) are compared in figure 4.6. The best fit is obtained for a nonlinear refractive index $n_2 = 1.6 \times 10^{-14}$ cm^2 W^{-1} and $L_c = 4.37$ mm.

4.3.1 The nonlinear coupler as time division demultiplexer

To test the AlGaAs integrated coupler as TDM demultiplexer (in digital communication sometimes referred to as a DEMUX) the high intensity control pulses and

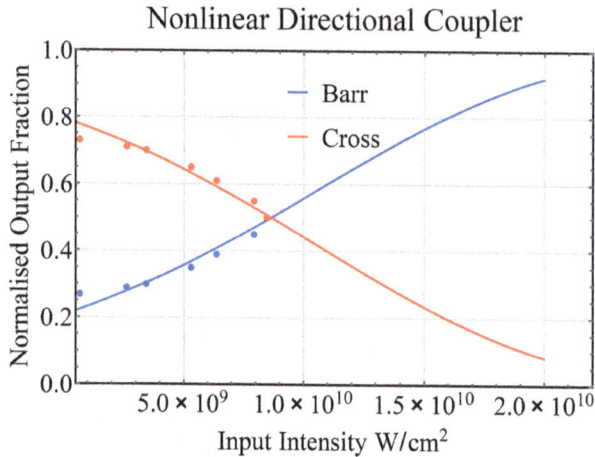

Figure 4.6. The figure shows the comparison of the nonlinear coupler performance with the theory. The red dots are experimental measurements of the output from the cross waveguide and the blue dots are measurements of the output from the bar waveguide. The measurements are normalised against the total throughput. The lines show the theoretical fit using the following parameters $n_2 = 1.6 \times 10^{-14}$ cm^2 W^{-1} and $L_c = 4.37$ mm. The notebook NLC Model.nb is used to fit the model to the data.

the low intensity 'data' pulses were derived from the same source, a mode-locked F-centre laser that produced 400 fs pulses at a wavelength of 1550 nm and at a repetition rate in the range 76 to 19 MHz [12]. To facilitate the separation of the control and data pulses at the output of the nonlinear coupler the pulses were inserted into the device with opposite polarisations. The data pulses were TM and the control pulses were polarised TE. At the output, a polariser was used to separate the control and data pulses. Both the control and data pulses were focused into the bar waveguide. The control pulse intensity in the bar waveguide was up to 35 GW cm^{-2}; that intensity was large enough to detune the coupling to the cross waveguide.

The results are illustrated in figure 4.7. It shows that when the control pulse is present in the bar waveguide the data pulse stays in the bar waveguide and when the control pulse is absent the data pulse moves over to the cross the waveguide. The control pulse 'shepherds' the data and keeps it in the bar waveguide.

The data rate here is limited by the repetition rate of the pulse source that is only a few 10s of MHz but, fundamentally, because the switch relies on bound electron nonlinear refraction the switching speed is fast and could switch at much higher data rates. Further, the switch is not limited by thermal considerations because neither the control nor data pulses are absorbed by either linear or nonlinear mechanisms.

A more efficient version of the nonlinear coupler was presented in [13] that used different wavelengths to separate data and control pulses, see figure 4.8.

As has been noted in chapter 3, the response of the bound electron nonlinear effect is around 115 attoseconds and so the picosecond pulses that are used to measure the switching are much longer than the response time. The low intensity

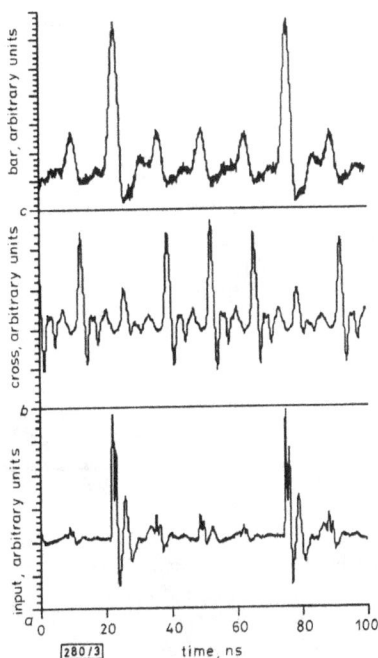

Figure 4.7. The time demultiplexing with a nonlinear coupler. When the control pulse is present in the bar waveguide the data pulse remains in the bar waveguide. The control pulse and the signal pulses are at the same wavelength but have orthogonal polarisation. The control pulse intensity in the waveguide is 35 GW cm^{-2}. Reproduced from [12] with permission of IET Publishing.

wings of the picosecond pulse do not have enough intensity to switch the device and the net effect is that only the central high intensity part of the pulse remains in the bar waveguide. The device therefore breaks up the pulse and this has been observed [14].

4.4 The Mach–Zehnder interferometer switch operating with n_2

4.4.1 Symmetric Mach–Zehnder interferometer (SMZI)

The integrated Mach–Zehnder interferometer presented in section 4.2 for the second order cascade device can also be used with the third order nonlinear refractive index. A single input waveguide is split at a Y junction into waveguides that form the arms of an interferometer. The arms recombine at a second Y junction and the phase relationship between the separated optical pulses determines the amplitude of the output of the device symmetric Mach–Zehnder interferometer (SMZI).

In figure 4.9 we show a symmetric Mach–Zehnder interferometer (SMZI) optical waveguide layout configured so that optical control pulses can alter the phase change in one arm of the interferometer so that the control pulses can switch the data pulse output of the interferometer [15].

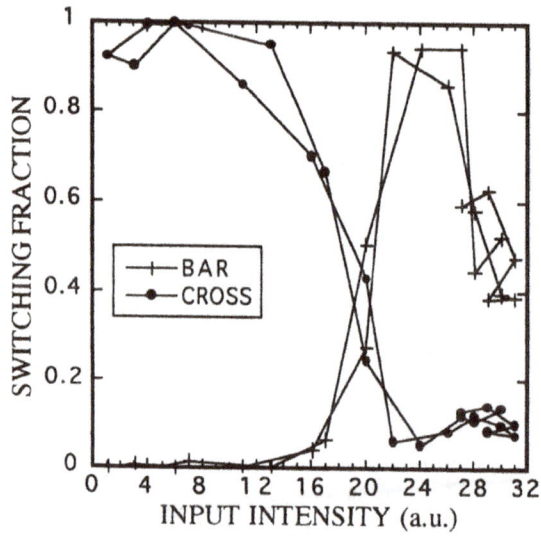

Figure 4.8. The switching fraction in a nonlinear coupler. The control and data pulses are at different wavelengths. © 1995 IEEE. Reprinted, with permission, from [13].

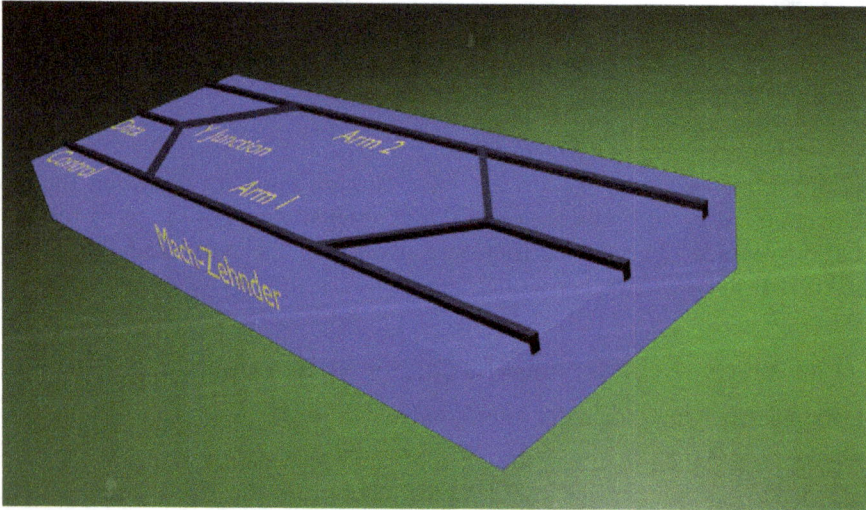

Figure 4.9. The optical waveguide layout of the integrated Mach–Zehnder interferometer with data and control optical pulse inputs for all-optical switching.

The phase in one arm of the interferometer is altered via the nonlinear refraction, n_2. If the intensity of the control pulse is I_{con} then the modulation of the data output, m, is given by:-

$$m = \cos^2(\Delta\theta/2) \qquad (4.8)$$

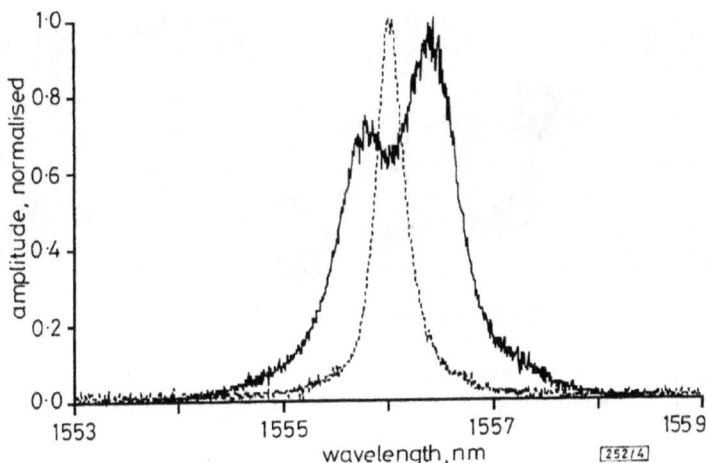

Figure 4.10. Spectral broadening of the high intensity control pulse in SMZI. The dotted line is the normalised spectrum of a low intensity optical pulse (40 MW cm^{-2}) that induces essentially zero phase change in the control arm of the interferometer and the solid line the normalised spectrum of a high intensity pulse (1.23 GW cm^{-2}) that induces $3\pi/2$ phase change in the control arm of the interferometer. Reproduced from [15] with permission of IET Publishing.

where

$$\Delta\theta = \left(\frac{2\pi}{\lambda}\right) I_{\text{con}} n_2 L \qquad (4.9)$$

L is the length of the interferometer arm and λ is the wavelength of the data pulse.

The SMZI integrated optics device was fabricated in AlGaAs and tested [15]. The device had arm lengths, L, of 15 mm and the core of the waveguide was $Al_{0.18}Ga_{0.82}As$ with a waveguide cross section of 15 μm^2. The control pulses, at $I_{\text{con}} = 0.6$ GW cm^{-2} were shown to modulate the data pulses by ~50% and that was consistent with a value of $n_2 = 1.3 \times 10^{-14}$ cm^{-2} W; the depth modulation was limited by the pulse break-up of the 10 ps pulses used to characterise the device.

There is an interesting aspect of the SMZI: the control pulse used to switch the device undergoes significant self-phase modulation, as illustrated in figure 4.10. The spectrum of the control pulse is measured after it has propagated in the control arm of the interferometer. At low intensity the spectrum is the same as the input spectrum, but at high intensities sufficient to induce a phase change to switch the device the spectrum has considerably broadened. This is due to self-phase modulation, which is a well-known third order nonlinear four-wave mixing process [16]—so after switching, the control pulse has more phase uncertainty. The price the system has paid for the switching operation is phase uncertainty in the control pulse. It relates to the quantum squeezed state effect, as predicted in [17].

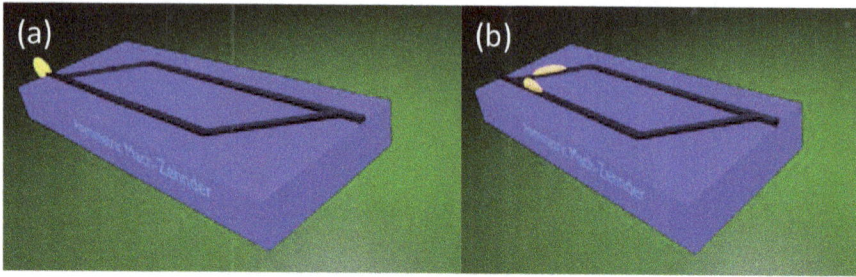

Figure 4.11. (a) The waveguide layout of the asymmetric Mach–Zehnder interferometer. (a) The yellow dot represents an optical pulse entering the input waveguide, (b) the light splits unevenly in the first asymmetric Y junction. The output of the device depends on the relative phase of the light in each arm as it recombines at the second Y junction. The relative phase is intensity dependent via the nonlinear refractive index.

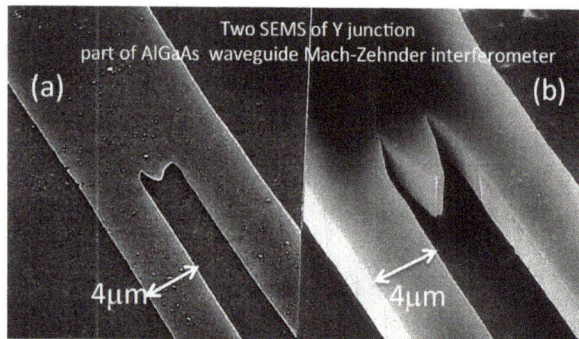

Figure 4.12. Figures (a) and (b) show two SEMs at different magnifications of the Y junction region of the AlGaAs asymmetric integrated Mach–Zehnder interferometer. The larger waveguide on the top left is split into two waveguides. The notch in the centre is due to imperfect fabrication.

4.4.2 Asymmetric Mach–Zehnder interferometer (AMZI)

The waveguide asymmetric Mach–Zehnder interferometer (AMZI) is illustrated in figure 4.11. It is a simple design that demonstrates the principle of Mach–Zehnder ultrafast all-optical switching [18]. In many ways it is the integrated optics version of an optical fibre device, the nonlinear loop mirror [4].

The device consists of an input waveguide section that guides the light to an asymmetric Y junction that splits the input intensity into two uneven parts. Figure 4.12 shows a scanning electron micrograph (SEM) of the Y junction. The split depends on the angle of the Y junction. The output from the device depends on the difference in the phase accumulated in each arm of the interferometer that in turn depends on the input power because of the power dependent refractive index. The uneven split in power is required so the power dependent phase shift is different in each arm of the interferometer. The phase difference at the second Y junction (near the output) is $\Delta\theta = \theta_1 - \theta_2$, where $\theta_{1,2}$ are the phase changes in the arms of the interferometer. If the power split between the two arms is as follows: δ in arm

AMZI transmission versus input power

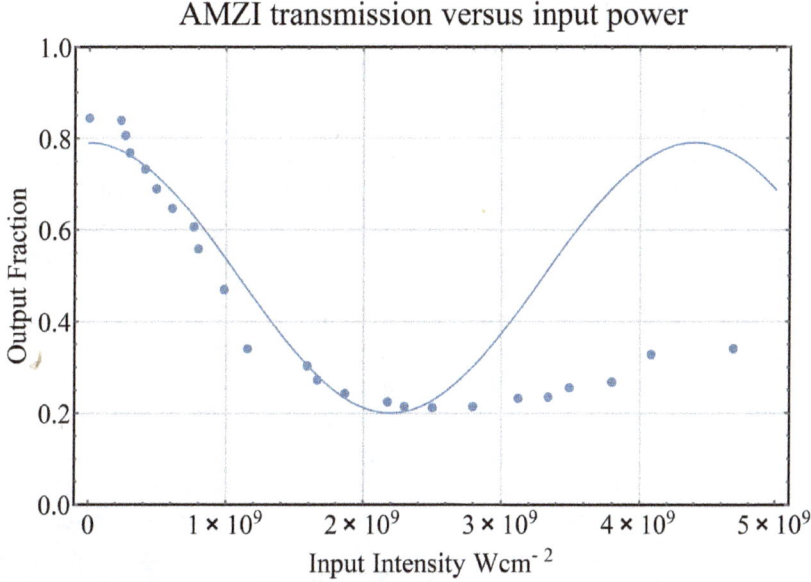

Figure 4.13. The figure shows the fit of equation (4.11) (solid line) to the data (solid circle) given in [18] for the asymmetric integrated Mach–Zehnder interferometer made with AlGaAs waveguides. The best fit is with $n_2 = 5 \times 10^{-14}\,\text{cm}^2\,\text{W}$. The fit is not good after the first null due to pulse break-up. The mathematica notebook AMZI Model.nb is used to fit the data.

1 and $1 - \delta$ in arm 2 (assuming no losses) then if we assume that $\Delta\theta$ is due entirely to the nonlinear refractive index, then $\Delta\theta$ is given by:

$$\Delta\theta = \frac{2\pi I_{in} n_2 l(1 - 2\delta)}{\lambda_0}, \tag{4.10}$$

where l is the length of an arm and λ_0 is the operating wavelength.

The transmission through the device, T, is given by:

$$T = 4\pi\delta(1 - \delta)\cos^2\left(\frac{\Delta\theta}{2} + \vartheta\right), \tag{4.11}$$

where ϑ is phase changes due to path differences perhaps introduced by fabrication imperfections.

An AMZI device was made in the AlGaAs waveguides [18]. The arms of the interferometer were 5 mm long and the power split at the Y junction $\delta = 0.18$.

The AMZI device was tested using high intensity, 330 fs pulses from a colour centre laser operating at a wavelength of 1520 nm. The results are plotted in figure 4.13 along with the model based on equation (4.11). The model does not take account of the fact that the low intensity wings of the 330 fs pulse are not switched off by the device plus in the nonlinear optical waveguides, self-phase modulation takes place that will tend to spread out the pulse in time. So because the theory is incomplete it does not describe the behaviour of the device at high intensities and after the first null the fit is not good.

4.5 Conclusions

As described in chapter 3, nonlinear optical effects associated with the second and third order nonlinearity can be used to produce an intensity dependent phase change in light propagating in optical waveguides. In this chapter we have described integrated optical devices in AlGaAs that make use of this intensity dependent phase change to achieve all-optical switching.

The cascade second order nonlinear effect can be used in an integrated Mach–Zehnder interferometer configuration that makes use of the feature that allows the sign of the nonlinear phase to be changed according to the sign of the phase mismatch. So the nonlinear phase change can be engineered to be equal but opposite in the two arms of the interferometer—the so-called push–pull configuration. A simple push–pull Mach–Zehnder interferometer is modelled with parameters appropriate for AlGaAs integrated optics device switching at optical communica-tion wavelengths. The input intensities required to switch the device of length around 1 cm are 2×10^6 W cm^{-2}, so that implies optical pulse energies of around 1×10^{-13} J (assuming pulse widths 300 fs and waveguide area 1.6×10^{-7} cm^2) and switching times in the attosecond region.

The nonlinear coherent coupler was one of the devices we described for the intensity dependent phase change induced by the third order nonlinear effect—the nonlinear refractive index, n_2. This integrated optical device consisted of two AlGaAs waveguides in close proximity so that there was overlap between the optical modes of the waveguides. In the linear regime (low power input) the light is coupled between the two waveguides and light in the input, bar, waveguide is coupled to the cross waveguides. This only happens when the waveguides have the same propagation constant. As the input power is increased the light intensity in the input waveguide changes the propagation constant via the n_2 effect and the bar waveguide has a different propagation constant from the cross waveguide, therefore, the power stays in the bar waveguide. Experiments on this device used the semiconductor alloy Al$_{0.18}$Ga$_{0.82}$As as the core of the waveguide—the alloy fraction was chosen to maximise n_2 while minimising β_2 (the two-photon absorption coefficient). An excellent fit between theory and observation was obtained for this device. The input intensities required to switch this nonlinear coupler device of length around 0.5 cm are 1×10^{10} W cm^{-2}, which implies optical pulse energies of around 1×10^{-9} J (assuming pulse widths 300 fs and waveguide area 1.6×10^{-7} cm^2) and switching times in the attosecond region. The high speed response of the device causes the pulses to break-up.

The capability of the nonlinear coupler as a demultiplexer was described and demonstrated by using high intensity, 400 fs pulses from an F-centre laser for data and control pulses that were orthogonally polarised. At the output of the device data pulses could be switched between the bar and cross waveguides by high intensity control pulses.

The SMZI used the nonlinear refractive index, n_2, to modulate data pulses and it required control pulse powers $I_{con} = 0.6$ GW cm^{-2} or control pulse energies $E_{con} = 1 \times 10^{-10}$ J (assuming pulse widths 10 ps and waveguide area 1.5×10^{-7}cm^2). As with the nonlinear coupler, the semiconductor alloy Al$_{0.18}$Ga$_{0.82}$As was used as the waveguide core. Further, it was observed that the control pulse underwent self-phase modulation.

The nonlinear integrated Mach-Zehnder was also configured as an asymmetric device. For the AMZI device of length around 0.5 cm, the input intensities required to switch are around 2×10^9 W cm^{-2}, which implies optical pulse energies of around 1×10^{-10} J (assuming pulse widths 300 fs and waveguide area 1.6×10^{-7} cm^2) and switching times in the attosecond region. The fit between theory and observation for this device was good at lower input intensities ($<2 \times 10^9$ W) but failed at higher intensities probably due a combination of pulse break-up and other nonlinear optical effects.

References

[1] Doran N J and Wood D 1988 Nonlinear-optical loop mirror *Opt. Lett.* **13** 56–8

[2] Wabnitz S and Eggleton B J 2015 *All-Optical Signal Processing* (Springer Series in Optical Sciences vol 194) (Berlin: Springer)

[3] Asobe M, Kanamori T and Kubodera K 1992 Ultrafast all-optical switching using highly nonlinear chalcogenide glass-fiber *IEEE Photonics Technol. Lett.* **4** 362–5

[4] Cotter D *et al* 1999 Nonlinear optics for high-speed digital information processing *Science* **286** 1523–8

[5] Islam M N 1992 *Ultrafast Fiber Switching Devices and Systems* (Cambridge: Cambridge University Press)

[6] Ironside C N, Aitchison J S and Arnold J M 1993 An all-optical switch employing the cascaded 2nd-order nonlinear effect *IEEE J. Quantum Electron.* **29** 2650–4

[7] Kanter G S *et al* 2001 Wavelength-selective pulsed all-optical switching based on cascaded second-order nonlinearity in a periodically poled lithium-niobate waveguide *IEEE Photon. Technol. Lett.* **13** 341–3

[8] Lee D L 1986 *Electromagentic Principles of Integrated Optics* (New York: Wiley)

[9] Hassan M T *et al* 2016 Optical attosecond pulses and tracking the nonlinear response of bound electrons *Nature* **530** 66

[10] Jensen S M 1982 The non-linear coherent coupler *IEEE J. Quantum Electron.* **18** 1580–3

[11] Aitchison J S *et al* 1991 Ultrafast all-optical switching in Al0.18Ga0.82As directional coupler in 1.55 micron spectral region *Electron. Lett.* **27** 1709–10

[12] Villeneuve A *et al* 1993 Demonstration of all-optical demultiplexing at 1555 nm with an AlGaAs directional coupler *Electron. Lett.* **29** 721–2

[13] Villeneuve A *et al* 1995 Efficient time-domain demultiplexing with separate signal and control wavelengths in an AlGaAs nonlinear directional coupler *IEEE J. Quantum Electron.* **31** 2165–72

[14] Villeneuve A *et al* 1994 Mode-locked and AlGaAs for nonlinear integrated-optics at 1.55 microns *Solid State Lasers Amplifiers Appl.* **2041** 153–64

[15] Bell J *et al* 1995 Demonstration of all-optical switching in a symmetrical Mach–Zehnder interferometer *Electron. Lett.* **31** 2095–7

[16] Boyd R W 2008 *Nonlinear Optics* (Amsterdam: Elsevier)
[17] Kitagawa M and Yamamoto Y 1986 Number-phase minimum uncertainty state with reduced number uncertainty in a Kerr nonlinear interferometer *Phys. Rev.* A **34** 3974–87
[18] Alhemyari K *et al* 1992 Ultrafast all-optical switching in GaAlAs integrated interferometer in 1.55 micron spectral region *Electron. Lett.* **28** 1090–2

Chapter 5

Conclusions

5.1 Discussion

We have covered second and third order optical nonlinearities and the related linear electro-optic effect, electroabsorption and electrorefraction and how they are applied in semiconductor integrated optic devices for switching and modulation with a strong emphasis on devices based on AlGaAs materials.

As a way of introducing all-optical switching we discussed electroabsorption, and electro-refractive effects that are currently used in optical communication systems for electrical to optical data conversion. The electro-optical effect is related to the second order nonlinear optical effect and the electroabsorption (and electrorefraction) effect is related to the third order optical nonlinearity. Both second and third order optical nonlinearities can be employed for all-optical switching.

From the theory of the cascaded second order nonlinear optical effect a push–pull Mach–Zehnder interferometer switch (the CasMZI device) can be designed which takes advantage of the ability to change the sign of the optical induced phase change according the phase mismatch condition. The device was modelled and with parameters appropriate for AlGaAs it was possible to design a Mach–Zehnder interferometer which is predicted to have an ultrafast response and low switching energy requirement compared to devices based on third order nonlinear refraction.

From the theory of the two-photon effect in semiconductors and in a variety of other materials we can design devices that use ultrafast optical nonlinearities involving bound electrons for all-optical switching. For optical communications wavelengths (around 1550 nm) the optimum design used $Al_{0.18}Ga_{0.82}As$ alloy as the core of the optical waveguides employed in integrated optical devices that turn induced optical phase changes into amplitude modulation. Nonlinear optical devices based on electro-optic device designs have been fabricated and characterised, the NLC, SMZI and AMZI.

Also, if we combine the two-photon absorption theory with semiconductor laser diode theory it is possible to predict the two-photon gain response of semiconductor optical amplifiers, as measured by Nevet *et al* [1] in AlGaAs waveguides.

In figure 5.1 we put the work discussed here on all-optical switching in AlGaAs waveguides in the context of the other work on digital photonic switching for optical communications systems that could work up to 100 GB s^{-1}. We have adopted the same metrics as previous workers for this comparison [2]. The results for NLC, SMZI and AMZI are from experimental results and the CasMZi numbers come from a model.

The switching landscape presented in figure 5.1, as with all metrics (switching energy and device footprint), comes with some built-in assumptions and needs to be used carefully. Really, these metrics are set up with incumbent CMOS switching technology in mind. The all-optical switch technologies, PPLN, HNLF and AlGaAs are intrinsically very fast and the limitations are the optical pulses used to measure the effect. The switching energy, $E_s \sim P_p \Delta t$, where P_p is the peak power and Δt is the pulse width. So the nonlinear optical effects are peak power dependent and then switch energy is therefore linearly dependent on the pulsewidth. As an example, take SMZI; it was measured using 10 ps pulses but the device would also work with the same peak power but a pulse width of 100 fs so then the switching energy, $E_s \sim 10^{-11}$ J not 10^{-9} J.

The SOA technology is more difficult to compare because although the device is all-optical and fast it has to be constantly supplied with energy to maintain the optical amplification and optical nonlinearity.

The switching energy metric assumes that the switching pulse energy is converted into heat and cannot be reused to operate another switch—as is the case with CMOS. But with the all-optical switches, the control pulses are not converted into heat and it may be possible to reuse the control pulses for another switching

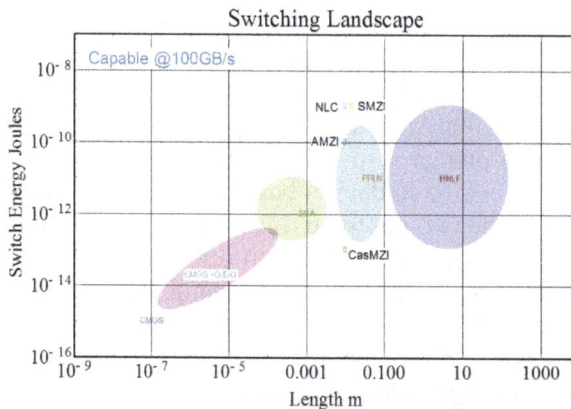

Figure 5.1. Summary of where the all-optical switching devices based on AlGaAs integrated optic devices fit in the switching landscape. NLC nonlinear coupler; SMZI the symmetric Mach–Zehnder interferometer; AMZI asymmetric Mach–Zehnder interferometer; CasMZI second order cascade push–pull Mach–Zehnder interferometer.

operation. This would require another type of switching architecture and indeed would not really be switching as we currently know it, and it presupposes some technology that is either embryonic or as yet non-existent, such as a good way of storing optical pulses in a recirculating optical memory.

The CasMZI device has a different status to the other technologies mapped out in figure 5.1, it is only modelled and has not yet been experimentally realised although similar bulk devices have been experimentally characterised [3].

The NLC shown in figure 5.1 has been taken all the way to a device that acted as a DEMUX.

5.2 Quantum effects

Although in the all-optical switching devices the energy in the switch pulse is not converted into heat, the all-optical switching pulse does not remain unchanged. For example, if we take the SMZI device, the spectrum of the pulse is altered by self-phase modulation, see [4]. So the price paid by the system for the switching operation is not heat but there is some entropy introduced by the phase uncertainty in the spread of control pulse spectrum. This may be fundamental and related to a quantum effect known as the quantum non-demolition effect [5].

To understand how the quantum non-demolition effect is relevant for all-optical switching we can look at the SMZI (section 4.4.1)—although the argument presented here is true for all the all-optical switches. In the SMZI the data pulse is switched by the control pulse present in the bar waveguide but the control pulse is not absorbed; so another way of regarding this is that measurement of the number of photons in the control pulse has been made by the data pulse without any of the control pulse photons being destroyed. That is a non-demolition measurement. According to one version of the Heisenberg uncertainty principle, the uncertainty in photon number, Δn, and the phase uncertainty, $\Delta \phi$, are related according to $\Delta n \Delta \phi > 1/2$. So in the SMZI the data pulse has reduced the uncertainty in the control pulse amplitude, Δn, thereby resulting in the uncertainty of the control pulse phase $\Delta \phi$ increasing; thus we observe the control pulse spectrum spreading—see figure 4.10.

The first of the optical quantum non-demolition experiments utilised the Kerr effect in optical fibres in the classical nonlinear optics part of figure 1.2 [6]; however, according to [5], true quantum non-demolition effects require being in the quantum photon–photon nonlinear optics part of figure 1.2.

The sort of self-phase modulation induced spectral broadening shown in figure 4.10 is characteristic of a number squeezed quantum state as predicted by [7].

5.3 Future prospects

It is clear from figure 5.1, that even with the O/E/O overhead, CMOS is still the dominant switching technology for practical optical communication systems. Although the all-optical switching of the PPLN, HNLF and AlGaAs does not impose a heat management problem, there is the problem of self-phase modulation of the control pulse. It is possible to handle that with dispersion management and arrange for the control pulse to have soliton-type properties where the self-phase modulation and the dispersion compensate, and that results in a pulse that does not change shape. However, it seems likely that the price of the entropy that is introduced to the control pulse will be paid for in some form of extra energy that has to be supplied to restore the control pulse so that it can reused.

Actually, from the theory presented in this book there is good reason to believe, as is mapped in figure 1.2, in the classical nonlinear optics region, for wavelengths around 1550 nm, that really $Al_{0.18}Ga_{0.82}As$ is about as good as it gets in terms of size of device and speed of the effect with minimum nonlinear losses. As for the linear losses—they are limited by fabrication tolerances.

Another approach is to try to move towards the top left hand corner of figure 1.2 and utilise quantum photon–photon nonlinear effects; that essentially means radically increasing the strength of the nonlinear effect without introducing a large increase in switching times. As outlined in [8], the strength of the nonlinear optical effects can be massively increased by employing cold atomic vapour in combination with resonant structures. There has been some progress in the engineering of cold atomic vapours in terms of making them compact and manufacturable [9], but utilising them in practical optical communication systems and achieving high speed combined with a large nonlinear optical effect remains an open challenge. However, aspects of that challenge are currently being tackled, albeit with a different agenda, by the quantum computing community, in which case the entities transporting and processing information may not be bits but qubits requiring a lot more sophisticated formats and data protocols compared to simple OOK.

References

[1] Nevet A, Hayat A and Orenstein M 2010 Measurement of optical two-photon gain in electrically pumped AlGaAs at room temperature *Phys. Rev. Lett.* **104** 4

[2] Hinton K *et al* 2008 Switching energy and device size limits on digital photonic signal processing technologies *IEEE J. Sel. Top. Quantum Electron.* **14** 938–45

[3] Kanter G S *et al* 2001 Wavelength-selective pulsed all-optical switching based on cascaded second-order nonlinearity in a periodically poled lithium-niobate waveguide *IEEE Photon. Technol. Lett.* **13** 341–3

[4] Bell J *et al* 1995 Demonstration of all-optical switching in a symmetrical Mach–Zehnder interferometer *Electron. Lett.* **31** 2095–7

[5] Grangier P, Levenson J A and Poizat J P 1998 Quantum non-demolition measurements in optics *Nature* **396** 537–42

[6] Friberg S R, Machida S and Yamamoto Y 1992 Quantum-nondemolition measurement of the photon number of an optical soliton *Phys. Rev. Lett.* **69** 3165–8

[7] Kitagawa M and Yamamoto Y 1986 Number-phase minimum uncertainty state with reduced number uncertainty in a Kerr nonlinear interferometer *Phys. Rev. A* **34** 3974–87

[8] Chang D E, Vuletic V and Lukin M D 2014 Quantum nonlinear optics—photon by photon *Nat. Photon.* **8** 685–94

[9] Nshii C C *et al* 2013 A surface-patterned chip as a strong source of ultracold atoms for quantum technologies *Nat. Nanotechnol.* **8** 321–4

Chapter 6

Mathematica programs appendix

The Mathematica notebooks are included because they were actually used in a number of cases in the work discussed above to help with the design of devices, and so they can help the reader understand the design process and thereby guide a reader who wants to design their own devices. The notebooks are meant to provide a deeper insight into how the physics of the devices is used to achieve the engineering objectives associated with high-speed switching of light.

6.1 What is Mathematica?

It is a mathematical symbolic computation program, sometimes termed a computer algebra system or program. It is a proprietorial (it is not free of charge) program produced by Wolfram that runs with a variety of operating systems. It has a notebook graphical user interface (GUI) that interfaces with a kernel that does the computation. With notebook, GUI equations can be entered pretty much as they look in the papers and textbooks, and publication quality graphics can be produced. Wolfram does have a website where programs can be run for free [1]. Mathematica notebooks can be converted into a computable document format (CDF) [2], and with the aid of a freely downloaded reader, CDF documents will run on all the common computer operating systems.

6.2 Why Mathematica?

It is the author's preference, partly because of legacy software that was used to actually design some the devices discussed in the book. The notebook format is particularly convenient for design, allowing the designer to play around with ideas to achieve major and small tweaks that can improve device performance. Mathematica is not perfect and can be inscrutable when things go wrong. It can be frustrating to find bugs because it does not provide any easy obvious way to dismantle the seamless notebook interface.

6.3 The notebooks

The notebooks all come with a general note of warning; they give estimates that need to be checked against measurement.

6.3.1 Chapter 1 notebooks

1. The notebook Refractive index for AlGaAs.cdf: plots the refractive index dispersion of $Al_xGa_{1-x}As$ refractive index as a function of wavelength (the refractive index dispersion) for wavelengths longer than the GaAs band-gap wavelength (900 nm) and as a function of x, the Al fraction in the alloy. The value of the refractive index obtained from this notebook can be used in the design of $Al_xGa_{1-x}As$ optical waveguides.

2. The notebook Waveguide Effective Index Channel.nb: uses the effective index method to calculate the optical intensity distribution for modes in channel dielectric waveguides. It uses Refractive index for AlGaAs.cdf to calculate the refractive index for different Al fractions. See also the notebook Waveguide Effective index Slab.nb.

6.3.2 Chapter 2 notebooks

3. The notebook FK&FOM@telecoms wavelengths.cdf: calculates electroabsorption in a bulk semiconductor, the Franz–Keldysh effect, for the InAlGaAs alloy. The alloy fraction is set so that the band-gap energy is close to photon energies commonly used in optical fibre telecommunication systems. Electroabsorption is used in optical communication systems to encode light with digital information often using optical waveguide formats in electro-absorption modulators (EAM).

The modulation figure of merit (FOM) for the InAlGaAs alloy is calculated for the device as it switches between two electric fields, F1 and F2. The electrical F1 represents the electric field present in the device due to the p-i-n junction before it reversed biased by the digital signal voltage producing the field F2. The digital signal represents the digital information to be encoded on the light beam. When field F2 is present the device is switched off and does not transmit light. The figure of merit, $\Delta\alpha/\alpha$ ratio, is a good measure of a low loss, high contrast on–off keying (OOK) EAM and it is plotted as a function of wavelength.

The magnitude of F1 also represents other effects such as the smearing of the band-edge absorption due to alloy fluctuations.

6.3.3 Chapter 3 notebooks

4. The notebook Cascaded 2nd order.nb: calculates and plots the exchange length, the depletion of the fundamental and the nonlinear phase change for the cascaded second order nonlinearity in GaAs. It uses the treatment given in [3].

5. The notebook Phase Missmatch.cdf: calculates and plots the effect of phase mismatch on the second order cascade nonlinear phase shift. The effect of the sign of the phase mismatch on the sign of the nonlinear phase shift is illustrated.
6. The notebook Two Photon absorption and n2 AlGaAs dispersion.nb: calculates and plots β_2 and n_2 for $Al_xGa_{1-x}As$ for optical telecommunication wavelengths as a function of the Al fraction, x.
7. The notebook Dispersion of n2 for AlGaAs.cdf: calculates and plots the nonlinear refractive index, n_2, of the semiconductor alloy $Al_xGa_{1-x}As$ for optical telecommunication wavelengths as a function of the Al fraction, x.
8. The notebook Figure of Merit dispersion.nb: calculates and plots the wavelength dependence of all-optical switching figure of merit for $Al_xGa_{1-x}As$ for values of x. The figure of merit wavelength dependence is the ratio $n_2 : \beta_2$ dispersion.

6.3.4 Chapter 4 notebooks

9. The notebook Cascade Push–Pull Model.nb: calculates and plots the Exchange length, nonlinear phase change and response of the cascade push–pull switch.
10. The notebook NLC Model.nb: calculates and plots a fit to the data presented in [4] to the model for the nonlinear coupler given in chapter 4.
11. The notebook AMZI Model.nb: calculates and plots a fit to the data presented in the model for asymmetric Mach–Zehnder interfometer given in chapter 4.

References

[1] Wolfram 2017 Available from https://www.wolfram.com/programming-lab/
[2] Wolfram *Computable Documant Format*, available from https://www.wolfram.com/cdf/
[3] Ironside C N, Aitchison J S and Arnold J M 1993 An all-optical switch employing the cascaded 2nd-order nonlinear effect *IEEE J. Quantum Electron.* **29** 2650–4
[4] Aitchison J S *et al* 1991 Ultrafast all-optical switching in $Al_{0.18}Ga_{0.82}As$ directional coupler in 1.55 micron spectral region *Electron. Lett.* **27** 1709–10

www.ingramcontent.com/pod-product-compliance
Lightning Source LLC
Chambersburg PA
CBHW082112210326
41599CB00033B/6675